Broadband Society and Generational Changes

Participation in Broadband Society

Edited by Leopoldina Fortunati / Julian Gebhardt / Jane Vincent

Volume 5

Frankfurt am Main · Berlin · Bern · Bruxelles · New York · Oxford · Wien

Fausto Colombo / Leopoldina Fortunati (eds.)

Broadband Society and Generational Changes

PETER LANG
Internationaler Verlag der Wissenschaften

Bibliographic Information published by the Deutsche Nationalbibliothek
The Deutsche Nationalbibliothek lists this publication in the Deutsche Nationalbibliografie; detailed bibliographic data is available in the internet at http://dnb.d-nb.de.

ISSN 1867-044X
ISBN 978-3-631-60419-9
© Peter Lang GmbH
Internationaler Verlag der Wissenschaften
Frankfurt am Main 2011
All rights reserved.

All parts of this publication are protected by copyright. Any utilisation outside the strict limits of the copyright law, without the permission of the publisher, is forbidden and liable to prosecution. This applies in particular to reproductions, translations, microfilming, and storage and processing in electronic retrieval systems.

www.peterlang.de

Acknowledgements

The editors wish to thank Andrea Cuman, Elizabeth Moffatt, Matteo Stefanelli and Silvia Tarassi for their helpful contribution in the editing and proofreading of this volume.

The editors wish to thank Peter Lang AG, Berlin for their support in establishing a new series on new ICT and society called Participation in Broadband Society.

This publication is supported by COST Office and their staff is acknowledged for their assistance together with the COST Action 298 Chair Bartolomeo Sapio and Vice Chair Tomaz Turk. The members of COST Action 298 Working Group "Humans as e-actors" are thanked for their support during the production of this book. These are: Julian Gebhardt, Hajo Greif, Amparo Lasen, Claire Lobet-Maris, Andraž Petrovčič, Lilia Raycheva, Panayiota Tsatsou, Olga Vershinskaya and Jane Vincent.

⊂cost

COST – the acronym for European COoperation in the field of Scientific and Technical Research – is the oldest and widest European intergovernmental network for cooperation in research. Established by the Ministerial Conference in November 1971, COST is presently used by the scientific communities of 35 European countries to cooperate in common research projects supported by national funds. Web: www.cost.esf.org

 ESF provides the COST Office through an EC contract

 COST is supported by the EU RTD Framework programme

 COST 298 – Participation in the Broadband Society

The present publication is funded also by MIUR (Italian Ministry of University) as dissemination activity related to the Research Project of National Interest (PRIN 2006) *Media and generations in Italian Society*

Table of contents

Acknowledgments — 5

Fausto Colombo, Leopoldina Fortunati,
Introduction. Broadband, Media and Generational Approach: a New Starting Point? — 9

Part 1: Generational Changes — 17

Fausto Colombo,
The Long Wave of Generations — 19

Michael Corsten,
Media as the Historical New for Young Generations — 37

Piermarco Aroldi,
Generational Belonging Between Media Audiences and ICT Users — 51

Jukka Kortti,
The Problem of Generation and Media History — 69

Part 2: Broadband Generations — 95

Marriann Hardey,
ICTs and Generations– Constantly Connected Social Lives — 97

Giovanni Boccia Artieri,
Generational "We Sense" in the Networked Space. User Generated Representation of the Youngest Generation — 109

Andra Siibak,
Online Peer Culture and Interpretive Reproduction on the Social Networking Site Profiles of the Tweens — 121

Mutlu Binark, Günseli Bayraktutan Sütcü,
Usage Patterns of New Media by Turkish New Middle Class Young People — 133

Ariela Mortara,
Generations and Media Fruition of Social Networks — 149

Marco Centorrino,
The Image of the "Digital Native" and the Generation Gap — 159

Matteo Treleani,
The Access to Memory in Video Archives On-Line.
Generational Roles on YouTube and Ina.fr — 173

Agnese Vellar,
"Lost" (And Found) in Transculturation. The Italian Networked
Collectivism of US TV Series and Fansubbing Performances — 187

Leopoldina Fortunati,
Digital Native Generations and the New Media — 201

Vesna Dolničar, Sonja Müller and Marco Santi,
Designing Technologies for Older People: a User-Driven
Research Approach for the SOPRANO Project — 221

Alberta Contarello, Mauro Sarrica and Diego Romaioli,
Ageing in a Broadband Society. An Exploration on ICTs,
Emotional Experience and Social Well-being within a Social
Representation Perspective — 247

Eugène Loos,
Generational Use of New Media and the (Ir)revelance of Age — 259

Chiara Carini, Ivana Pais,
Business Social Networks: an 'Age Levelling' Service? — 275

Tanja Oblak Črnič,
The Generational Gap and Diverse Roles of Computer
Technology: The Case of Slovenian Households — 289

Authors — 305

Index — 313

Fausto Colombo, Leopoldina Fortunati

Introduction. Broadband, Media and Generational Approach: a New Starting Point?

The role generations play in accepting and shaping digital technologies, and possibly vice versa, is a particularly significant issue in current studies. For the first time in the academic debate, this volume tries to address the relations between media and generations. One of the main aims of this book is to outline the theoretical debate on the general theme of the relationship between generations and media, focusing especially on the media system as a whole after the spread of broadband.

This introduction doesn't want to deal with the debate over generations neither in its classic version that has accompanied the birth of human sciences, nor in its latest developments. This synthesis will be offered in the first part of the volume by some authors. Here some indications regarding origins and specific contents of each paper will be outlined.

This collection of papers is the result of a long process of networking and cultural exchange between scholars in Europe and the selected works were first presented at two major international events: one, ICT and Generations, was held at the conference entitled The Good, The Bad and The Challenging, organised by COST Action 298 in Copenhagen (13-15 May 2009) and the other, Media + Generations, took place in Milan on 11 and 12 September 2009 and was funded by the Italian Ministry of University and Scientific Research.

The first part of this two-part book is entitled *Generational Changes* and addresses the main issues concerning the theoretical debate on generations. Fausto Colombo's essay in particular, *The Long Wave of Generations*, re-visits recent work on the definition of generation from such disciplines as history, demography, marketing and sociology, putting forward some criticism of the dominant marketing model and suggesting a perspective that distinguishes sociological viewpoints: firstly regularity vs. exception and secondly, external look vs internal look.

In *Media as the "Historical New" for Young Generations*, Michael Corsten starts by outlining some basic assumptions in Karl Mannheim's concept of generations: young adults born in the same period would be especially sensitive to the 'Historical New' during their adolescence. The focus of Mannheim's was on political thought styles and world views, but nowadays it might be the case that the political dimension is not as relevant as it once was. Therefore, mass media forms and usages might be more important in the way contemporary young adults relate to public life and to the political dimension. The shifts of the 'Historical New', in the way Mannheim puts it, would then become visible via the shifts in mass communication media.

In *Generational Belonging Between Media Audiences and ICT Users*, Piermarco Aroldi presents some ideas about the relation between ICTs and the generations of their users. The main questions are firstly, whether, and if so, how, media contribute to the construction of generational identities; secondly, whether, and if so, how generational belonging affects media usage and sense-making in the everyday practices of media consumption; and finally, what the differences introduced by ICTs in this relation between "media" and "generations" are. After the proposal of a model of the generation-building process based on a definition of "generation" in agreement with a review of the sociological literature, the article describes the roles played by the media in this model and in different theoretical approaches and finally discusses the new developments introduced by ICTs.

The essay *The Problem of Generation and Media History*, written by Jukka Kortti, questions or at least remoulds and defines the sociological ideas of the generation in historical studies. The interest will be given on one hand to the role media play in creating generational consciousness – actively and reflectively or more in subconscious level. On the other hand how media shape the image of a generation, or how the discourse of a generation is realised through media. The examples are picked out from the Finnish media history, both press and broadcasting history, but particularly from the Finnish student magazine Ylioppilaslehti. The samples concentrate on the periods from 1920s to 1930s on one hand and from 1960s to 1970s (particularly 1968) on the other. Both periods were particularly interesting in the context of modernism, which is crucial for the theories of sociological generation.

The second part of the book is entitled *Broadband Generations* and gathers together a selection of research at international level, with particular although not exclusive attention being paid to the European context. It will be clear that, apart from the methodologies used, the research adopted three different approaches.

The first approach, used by the majority of the contributors, is the analysis of "domestication" processes taking place in younger generations - children, teenagers and young adults.

In *ICTS and Generations Constantly Connected Social Lives*, Mariann Hardey constantly connects social lives and delineates the issue of digital social networks and associated media in the lives of young people - primarily middle class young people - who have grown up in technology-rich environments. Others have identified this group as 'Generation Y', the 'Google Generation', the 'Internet Generation', the 'Facebook Generation', the 'Digital Generation' and the 'e-Gen', in attempts to capture the particular lifestyles and habits of relatively young consumers who have grown up in the Information Age. Within this culture, Web 2.0 applications - in particular Social Networking Sites (SNSs) such as Facebook - are seen as essential 'pivots' for the media-led lives of this generation. Drawing on the analysis of ethnographic data from the United Kingdom and Australia this paper illustrates how social media resources have become embedded into everyday social practices. Thus, individuals are in effect 'always connected' and are seen to be 'always on' and engaged with others converging material and digital lives. As a consequence such mediated interactions have given rise to, or are exposed as, new dilemmas and social opportunities when connecting to others.

Generational "We Sense" in the Networked Space. User Generated Representation of the Youngest Generation, by Giovanni Boccia Artieri, presents results of a research project about how young people generate generational narratives from mass media culture on the networked space. Weblogs, and the entire Web 2.0, seem to be a feasible place from which to observe whether and how generational discourses emerge, and whether they arise around specific media content or media topics which can trigger the reflexive process. Media products, e.g. novels, movies and TV shows, affect a broad process leading to the creation of a shared set of meanings and a shared sense of belonging, or what can be called a generational 'we sense'. The research observes not only how young people perceive their media representation but also how they are able to use those representations as components of their own personal narration.

Andra Siibak, in *Online Peer Culture and Interpretive Reproduction on the Social Networking Site Profiles of the Tweens*, presents a study focused on textual parts of the profiles of 11 and 12-year-old users of Rate, the most popular SNS in Estonia (www.rate.ee). Content analysis was done of 20 profiles of youngsters to discover what kind of communities children belonged to as well as what kind of hobbies, interests and tastes they proclaimed to have. Children were often engaged in socio-dramatic role-play while compiling their self presentations on the profiles. For instance, they often joined communities that could create a feeling of empowerment and control by them taking on adult roles. Nevertheless, children did not only imitate aspects of the adult world directly in their identity performances. Children were also involved in interpretive reproduction in order to appropriate the information gained from the adult culture and to produce their own creative version of peer cultures that would satisfy the needs of their peer world.

Mutlu Binark and Günseli Bayraktutan Sütcü (*The Usage Patterns of New Media by Turkish New Middle Class Young People*), analyze various usage patterns of the new media in everyday life by new young middle-class Turkish people. The research concentrates on how these people use the new media to mobilise space, transfer their offline social relations to online ones, experience two-dimensional parental control, and construct a communication landscape by combining various new media. Throughout this micro-scale analysis, the paper locates the usage of new media within a larger context of consumer culture, and argues that the new middle-class young people, potentially civilised citizens, are learned as indifferent, egocentric and consumption-oriented individuals.

Generations and Media Fruition of Social Networks, by Ariela Mortara and Enrico Marchetto, focuses on the typical use of social networks made by younger web surfers. These constantly connected individuals are always in touch – anywhere, at any time – but they are not the only subscribers to social network sites. Indeed, although, as some research highlights, the digital divide still separates younger people from the not so young, recent data underline that the average age of social network users is rising. Using recent research, the authors show that among Italian social network users it is possible to outline different ways the networks are used based on the age of subscribers and in relation to numerous variables: the level of technological skill, the way the Internet is accessed and the motivation driving people to open an account on a social network. The authors explorative research was conducted on quantitative data collected through participative research on social networks sponsored by an Italian research company.

In *The Image of the "Digital Native" and the Generation Gap*, Marco Centorrino presents a study of groups of young people or individuals in the 12 to 16-year-old age group which goes beyond the currently dominant images that seem to revolve around bullying, a term which covers all types of acts of prevarication and/or violence. The study has successfully shown how these 'digital natives' are multi-taskers, theme-oriented, also multi-networked, and act in networks of social relationships they establish not only in daily life but also in the sphere of virtual reality. They also demonstrate marked entrepreneurial abilities, exploiting all their internet tools in a context - that of the large user numbers of the internet - where they demonstrate their skill in moving with great agility, combining internet interaction with everyday life.

The Internet can be seen as an audiovisual resource that constitutes, in the view of researchers, a 'real' collective memory. If we consider the role of the younger generations in the use of new media, and the significant influence these new forms of knowledge have on young people, it is useful to examine the ways in which the web addresses these users and gives them access to this audiovisual memory. And this is the starting point of Matteo Treleani's essay *The Access to Memory in Video Archives. On-Line Generational Roles on YouTube and Ina.fr*.

Two websites demonstrate a different approach to this question: Ina.fr, the online database of Institut National de l'Audiovisuel, a publicly funded French enterprise with a commercial slant; and YouTube, the largest container of videos on the web. 'Surfing' these archives is tantamount to 'surfing' the collective memory. The intention of building an audiovisual memory online has been explicitly put forth by INA (whose slogan was, until 2008, "we build the future of your memory") and it is implicit in YouTube, a site that functions as a time capsule with an infinite quantity of television programmes archived by users.

Agnese Vellar's *"Lost" (And Found) in Transculturation. the Italian Networked Collectivism of US TV Series and Fansubbing Performances* is an investigation into how Italian television audiences participate in networked publics. The author conducted ethnographic research on the networked collectivism that emerged around US TV series, using the case of the community of Italian Subs Addicted (ItaSA), a organization of fans who produce amateur subtitles, or fansubs, for US TV series. The members of the staff are young adults who work in teams to produce subs without expecting financial reward. Their products are consumed by a young audience which is not satisfied with Italian national television dubbed versions of the US TV series which are broadcast long after its US distribution. The essay describes how members of Age Cohort 1979-1991 adopted and adapted technologies and media content to fulfill spectatorship needs and thereby participated in the construction of a generational imagined community in the networked publics.

A revision of the knowledge of the so-called digital native generation is the aim of the Leopoldina Fortunati's contribution: *Digital native generations and the new media*. The author draws data from two separate research projects: one in 2008 which investigated the relationship between social participation and new media (namely computers/internet and mobile phones) and one in 2010 which investigated the relationship between the second digital native generation and new media. The research shows that a second digital native generation is emerging, which is made up of the younger brothers and sisters of the first digital native generation. This second generation seems to emphasise the characteristics of the first generation.

The second of the approaches previously mentioned concerns the study of mature generations, and even generations of the elderly, and their effort to approach and understand in their own way new digital technologies and particularly the Internet. It is a less popular approach, but nevertheless useful because it will be very significant in the future when today's mature generations who are used to innovation and do not want to feel outdated eventually become old.

From this perspective, in *Designing Technologies for Older People: a User-Driven Research Approach for the SOPRANO Project*, Vesna Dolničar, Sonja Müller, Marco Santi present a user-oriented research approach and set out the results of a needs assessment and requirements analysis carried out within the SOPRANO project (http://www.soprano-ip.org/). Service Oriented Programmable smArt enviroNments for Older Europeans (SOPRANO) is a European Commission 6th Framework Programme integrated project which started in January 2007 and will go on until October 2010. SOPRANO is a consortium of commercial companies, service providers and research institutes with over 20 partners from Canada, Germany, Greece, Ireland, the Netherlands, Slovenia,

Spain and the UK. The project's main goal is to design and develop highly innovative, context-aware, smart services with natural and comfortable interfaces for frail and disabled older people at affordable prices and which meet the requirements of users, their families and their care providers. Contemporary societies, as the author aims to outline, are facing two striking trends: an ever larger ageing population and rapid diffusion of new technologies.

The research presented shows that a large segment of the growing number of older people in Europe can be offered services which could radically improve their quality of life but only if services and the ICT components supporting them are designed in a way that makes appropriate use of ambient intelligence and the new ability of software systems to communicate with users.

Alberta Contarello, Mauro Sarrica, Diego Romaioli (*Ageing in a Broadband Society. An Exploration on ICTs, Emotional Experience and Social Well-being within a Social Representation Perspective*) explore how ICTs are used and represented by elderly respondents as well as how ICTS are related to social well-being and emotional experience. Two venues of research are touched upon. One inquires into representations of ICTs, social well-being and emotional experience in elderly participants, and the other deepens our focus of attention with interviews with 'first generation' elderly users. The studies also more from a wider interest in the social representation of ageing, and in its connections with concepts such as active ageing and positive aging. Provisional results of the research show how important it is to take account of different components in the study of the relation between representations of ICTs and social psychological processes. Not simply age itself, but a rich variety of variables have to be considered when analysing the relationship between ageing and ICTs.

The third approach consists of comparative analysis of the relation of different generations to digital technologies and whether they privilege some over others and whether they adapt differently to the same set of technologies

In his paper *Generational use of new media and the (ir)relevance of age*, Eugène Loos focuses on netsurfing and asks: do older people really navigate websites differently to the way younger people do or are the differences within each generation, such as those mentioned above, greater than the differences between generations? To answer this, Loos presents data, including interviews, from an eye-tracking study carried out among 29 younger and 29 older users in the Netherlands. The paper highlights that, although differences in navigation behaviour are to some extent age-related, there is also evidence that differences within the group of older people are dependent on gender, educational background and frequency of internet use. What emerged was far more a digital spectrum rather than a digital divide between digital natives and digital immigrants.

Ivana Pais and Chiara Carini present a paper focused on the users of business social network services (*Business Social Networks: an 'Age Levelling' Ser-

vice?). It starts by referring to two main branches of BNS literature: one related to economic sociology, dealing with the role of social capital in business social networks; and the other related to social media studies, dealing with social network sites. The authors carried out an exploratory analysis of Milan IN, a non-profit association set up in 2005 to let members of LinkedIn living in Milan to physically meet up with each other.

A wide range of methods was used, such as participant observation, online questionnaires and semi-structured interviews, and the research succeeded in demonstrating how age influences whether a subject chooses to join a business social network but not how they use it.

Last, but not least, Tanya Oblak Črnič's *The Generational Gap and Diverse Roles of Computer Technology: The Case of Slovenian Households* analyzes the real behaviour of domestication of communication technologies in such typical intergenerational contexts as families – in this case Slovenian families.

Interviews with parents and their children show that the Internet is regarded as the most important dimension of the computer as a modern domestic appliance. As a consequence, computer technology plays an important role in the process of building all users' private spheres. Although to start with the home computer was considered primarily as "a business device" or "a toy", new meanings are attributed to it nowadays. The computer is not only "an investment for the future" or "the extension of parental responsibilities". It also has several controversial roles, such as that of an intruder" or "a destroyer of personal relationships" on the one hand, or that of "a comforter" and "a multi-tool" on the other, depending on the generation that the user belongs to.

In conclusion, we believe that this book goes a long way to providing a broad vision of all the various perspectives of generational approaches to research into media and digital technologies. It sets out several routes of enquiry which have adopted various research methods and which have provided up-to-date results, all within what we trust will be seen as a useful and acceptable theoretical paradigm.

First Part: Generational Changes

Fausto Colombo

The Long Wave of Generations

1 A Topical Issue

The generational issue is today highly topical. It is under debate, for instance, when talking about changes to the traditional welfare system, especially regarding the need to create a balance between older people, who are guaranteed, and younger who are not; or when talking about the renewal of the political class, in particular after the election of the new President of the United States, Barack Obama, as a member of a new round of statesmen outside the traditional – or old – establishment. The generational issue is again discussed to shed light on the problem of education, in an attempt to highlight difficulties in today's relationships between parents and children and between educators and students which can be traced back to their generational membership. The issue is again pivotal when dealing with the literature in which the sons of the great 1968 contesting generation reread accounts of their childhood experiences from a new perspective framed in the history and ideologies of the time[1]. The generational issue can be seen to be called into debate in public opinion and particularly in scientific thinking.

If we start by attending to the specific subject this volume deals with, we realise how common definitions of the most recent generations are often influenced by digital technologies as key elements of communication: the "Internet Generation", the "Nintendo Generation",the "Avatar Generation", to name a few, and perhaps very importantly, "Digital Natives" (Palfrey & Gasser, 2008; Prensky, 2001; Tapscot, 1998). As I will explain throughout this paper, these definitions cannot be considered valid for the purposes of explanation and could even be deemed dangerous because of their charm of use but lack of depth in accounting for evidence from society. However, they do have some advantages, inasmuch as they keep alive attention to generational issues and to

---

[1] In Italy, for example, there is a vast literature authored by the 1968 generation's sons. Just to make a few examples, I'll mention three types of discourse: the first two regard the autobiographies by terrorism victims, or viceversa, by the sons of the central characters in the active politics of those years. The third type instead, is made of the accounts of people which at the time of the transformations induced the 1968 generation were children.

the role played by the media: a relevant subject matter, albeit in a very different perspective from that behind the definitions mentioned above. But let's step back, to make clear the theoretical framework within which the debate about generations is currently taking place (Colombo, 2003).

First of all, it is useful to remember that the theoretical concept of "Generation" belongs to a very classical tradition of "sociological paradigms" (Aroldi, *Infra*). This tradition is still alive, and there are several authors actively contributing to redesign and update it. I refer in particular to Michael Corsten (of which this volume contains an essay of great interest) or to Edmunds and Turner (2002), whose definition can be taken as an excellent starting point: "An age cohort that comes to have social significance by virtue of constituting itself as a cultural identity".

The interest of sociology in this problem is not unexpected: talking about generations means talking about society, in its most profound and immediate aspects, precisely those aspects on which any sociological analysis is based on. First of all, the contrast between individual and society, between atom and galaxy. When talking about "Generations" we speak primarily of people as social actors with a life cycle that links them together with humanity and with the individuality of each person. However, we are also talking about the temporal dynamics that are the background to any social change, and that are substantiated in historical cycles, with their watershed, their characteristics, their overlaps, and where community is somewhat the principal player. Finally, we speak of those intermediate bodies or agencies which sociological analysis has often emphasised, such as family, school and the media. These are the places where the different generations meet, talk and sometimes clash among themselves; where a sense of continuity (the family), tradition (the school) and sharing (the media) is handed on and down and where, however, fractures that make generational differences evident occur. In sum, that sociology looks at generations as a peculiar form of social identity.

But before analysing this theme from a sociological point of view, I will briefly consider how some interesting issues close to disciplines such as demography, history and marketing might contribute to my discourse.

The starting point is surely demography, the discipline that can probably claim a privileged attention to generations. Here we should look at the etymological root of generation, which recalls the act of generating, of giving life to those who will follow us into the future. In demographic terms, time becomes essential and assumes through successive generations a specific form, a sort of undulating line shaped by the rhythm of births and life cycles. Of course, demographics has recently started looking at the role played by social factors (Di Giulio & Rosina, 2008) in the reconstruction of this wavy line, but it must be said that other elements are useful in understanding the specificity of individual generations, as will be seen below. For example, the weight, i.e. the number of people of a particle cohort age, is recognised as a pivotal factor contributing to the role generation may play in history. From this point of view, we should look at the relevance played by the consistency of generations in the aforementioned reassessment of the welfare state, where the advent of retirement age in very numerous cohorts undermines a system based on active work of smaller cohorts. Or, to begin fielding issues related to our interest, we should think about how the visibility and success of a technological or media innovation also depends on the population size of the generation as its privileged target. Thus, the European generation of Baby Boomers, by definition "heavy" in terms of population, has determined the success of such innovations as analogue media for music (45 and 33 rpm vynil records) or hi-fi more quickly and broadly than the younger generations with MP3s or iPods, which instead have been successful thanks to their intergenerationality. Furthermore, in terms of musical content, we should ask ourselves whether the success of rock, which still exceeds that of successive popular music genres, is not also due to the solid demographic of its original audience.

Another discipline fruitfully using the idea of generation is surely history. We are of course dealing with the most recent history, which ever more successfully interfaces with sociology (Straurs & Howe, 1997). The notion of generation appears in history as both causes and as effects of social change. To start from the second point of view (generations as effects of social change), it is clear that some specific historical conditions can influence and mark people born in a certain period. In Italy, for example, after the defeat of Caporetto-Kobarid (in 1917) those born in 1899, many of whose were not even eighteen, were called to arms to reinforce the troops at the front. They were the so-called "boys of '99": a generation at the forefront of the post-conflict period, either supporting or fighting against Fascism. According to an Italian writer, Alessandro Baricco (2005), this generation came out of the war completely transformed, and sought the opportunity to react to that tragic experience, up until the Second World War. In this sense, the so-called Millennials can be probably marked by the participation as an audience stunned by the fall of the Twins Towers (made possible globally by the media), and many other examples could be mentioned. However, the question which concerns the sociology of media is evident when looking at these two examples. Can events covered by the media be viewed in the same way as those that literally swept away a generation involved actively in history? It could be argued that for certain major historical events the difference is not always relevant, at least for some. Firstly, the "boys of '99" would have probably learned during trench warfare of the profound and truly historic meaning of a faraway event such as the assassination in Sarajevo (and covered with all the emphasis by the media of the time). And vice versa, for many young people from the US and from other allied countries, the attack on the Twin Towers has assumed a significance far more concrete in the battles of the wars in Afghanistan and then Iraq ... as if to say that there is a fine line between being a spectator and being an actor.

A different issue is the one concerning the so-called "active" generations, as that always evoked in historical and sociological studies in terms of the "exemplary generation" of participants in 1968. (Edmunds & Turner 2002) Certainly, this generation has some unique features: it was in every sense a global generation, not least thanks to the spread of the media and their ability to cover events from all over the world, from the United States and Mexico to China and Vietnam, Western Europe (France, Italy, Germany) and from the east (with the Prague Spring and its repression, see Berman, 1997). It was also a strong generation yearning for protagonism and willing to take history upon their shoulders. Moreover, following the great wave of 1968 (even though it is not always only because of it), Western societies were crossed by big changes in social values and in the recognition of common rights previously denied.

Also from this point of view, the role of the media should not be neglected, not only because, as mentioned before, media were partly and especially involved in showing young people around the world their similarities and their common membership, but also because those generations were using media, from stencils to newspapers, from cinema to radio, in a radical and innovative way.

The differences between the two models of generation (to use the notion of generation mentioned above, as causes or as effects of social change) are more logical than interpretative. Obviously there can be generations, as individuals, who appear to be more spectators rather than actors on the stage of history. But it is hard to imagine that actors can move without having their own idea of reality (that the media cooperate to form) (Silverstone, 2007) and, vice versa, spectators who somehow do not find themselves in the midst of change and do not participate in some way to it. Rather, as some historians remind us, the specific ways in which a generation interacts with others, and especially with the previous ones, should be investigated. One might well find that even some uses of media typical of a generation (for instance, for Italians baby boomers the invention of radio independent from state monopoly in the Seventies) are made possible both by the desire to challenge and overcome the media system of the previous generation, and by customs of use introduced by the generation of parents.

The reference to the issue of consumption helps to introduce another discipline that has been involved in tackling – in recent times – the generational problem: marketing. The first thing to point out is that the objects of investigations of this discipline are clearly different from those of the social sciences: if generation is primarily an identity for sociology and an active subjectivity for history, it is primarily a target for marketing (Smith & Clurman,

1997), i.e. it is a social segment engaged in coherent and unified behaviours (especially of consumption).

The definitions that marketing has applied to generations are therefore more operational than interpretative, and moreover they refer to different criteria. We can distinguish three of these criteria: a) definitions based on an supposed strong identity, determined by complex social factors shaping the generation (example: Baby Boomers); b) definitions based on a weak identity whose characteristics do not necessarily need to be taken on and whose behaviours, especially in terms of consumption (e.g. symptomatic: the so-called Generation X, see Craig & Bennet, 1997), are clear enough to be observed; c) deterministic definitions, in which the generation is characterised by a certain consumption, typically tied to the use of technological innovations (examples of this are the aforementioned Net generation, the Nintendo Generation, Digital Natives and so on, see Morice, 2002). This last definition can be regarded as mid-way between the first definition, which assumes the possibility of designating a generation from a recognizable feature of identity, and the second definition, with which it shares the centrality of consumption as an identifying element.

It is clear that the definitions mostly concerned with our standpoint are the most deterministic ones. I have already mentioned in the introduction that I do not find them useful to understand the complexity of generation, and it is now time to argue this position. The main reasons for my criticism to the generational marketing approach are substantially twofold: the pars destruens builds on the epistemological limits of these definitions, while the pars construens consists of the argumentation of my work, and will be the second part of this article.

Pars destruens: in the definition I have called "deterministic", it is taken for granted that the technology being used (or – as with the label "digital natives" – the complex technologies used) is the factor which distinguishes one generation from another. Usually the label is applied to the younger generations, who better accommodate and shape the supply coming from the market. The digital revolution, in particular, becomes a perfect watershed, because it changes the scenery of the available equipment by changing the previous ones and offering new ones. It also allows new social rituals, changing – at least apparently changing – the habits of privacy and sociability, the sense and use of time, and so on. Apparently it is a perfect example of how a technological revolution may drive a big social change (Tapscott, 1998).

Of course, such a position must firstly explain why this digital revolution particularly affects young people. The argument usually used is that young people are more malleable, because they are facing the training phase of their lives, and therefore they receive transformations in the technological landscape more naturally than adults, in terms of acceptability (they see new technologies as fashionable tools, essential if they are to be part of their peer group), in terms of literacy (they learn these technologies by use rather than through the traditional processes of schooling) and in terms of capacity for innovation and change (they are not tied by past habits). The argument has its value of course. And it is certainly true, for example, that the experiences of childhood shape individuals, and therefore also generations. However, when dealing with this lapalissian truth, experiences have to be understood n their the totality, which is obviously difficult to summarise from one aspect or another. And then, just reconsidering the example of the digital revolution, which of the arguments mentioned above cannot be applied to adults? Availability of innovation, for example, depends on a number of complex factors, such as level of education, property, and even those "moral economies", so rightly mentioned by Silverstone (1992).

Moreover, it could be argued that some generations are historically more open to new ideas than others. This is the case of Baby Boomers in all Western societies who experienced a period of great innovation during their youth (the mechanical and analogical revolution of technologies and individual consumption) and who throughout their life have seen a succession of innovations needed to support the changing markets from obsolescence of products and standards (think of the change of sound recordings from vinyl to CD to MP3). It is therefore probable that this generation of parents of the current so-called Digital Natives have positive attitudes towards innovation and that this attitude is encouraging both their use of innovation itself and the positive and encouraging attitude towards their children. Regarding then the willingness to acquire literacy by instinctive or mediated by peer-to-peer ways, it might be observed that there are traditional factors such as gender, which are recognised as significant differences in learning processes (for example, a certain superficial and instinctive speed of males as opposed to a greater reflexivity and also a greater tendency of women to completeness of use). Thus, the generational labels in this case should be made more complex by taking into account other factors.

Therefore I believe that many definitions or labels of generational marketing – although they might be useful for posing the question of the role of media in the construction of several generations – are often a very easy path, and must be taken by sociologists more as stimulus rather than as real interpretative categories.

I will now describe the complexity of the relationship between media and generations, summarizing what the sociological thinking has acquired and what empirical studies seem to reliably demonstrate.

2 Points of View

Firstly, it should be noted that in sociology the social fact is not separate from the observer's glance. Generations, like any social facts, change appearance depending on our approach. Now, it is very important to be aware of the fact that different aspects of generational phenomenon can be outlined by different gazes. Although each of these issues is part of the social observation, we must always keep in mind that the various portions observable cannot be added together in the hope of returning to the whole. We are doomed to see things "per speculum et in aenigmate", as St. Paul said. This does not mean that we haven't to keep on trying, and interpreting the portions of the phenomena our methods of investigation reveals to us.

In the case of generations, there are two main points of view: the first sees them as being included in the social continuum. From this point of view, generations are fantastic objects to question social changes and therefore as ways to see action over time. Obviously, we can represent this temporality as a regular pattern, or as a series of catastrophic changes, followed stages of adjustment of varying degrees of calmness.

Looking to future generations in terms of regularity may be – even in relation to our issue – very useful. An interesting metaphor is that of the succession of waves. While it is true, as demographics have gradually ascertained, that generations are not boxes of time, all of equal length, containing groups of people, it is also true that the succession of generations seems to be regular, although great events are capable of altering the rhythm. So we see that the succession of generations is like a series of waves, and may well be represented as such:

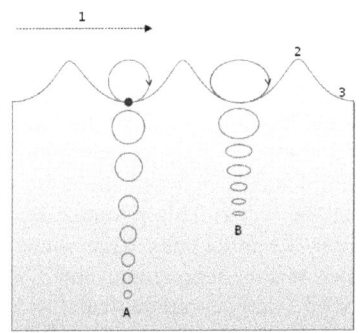

A = In deep water.
B = In shallow water. The elliptical movement of a surface particle becomes flatter with decreasing depth.
1 = Progression of wave
2 = Crest
3 = Trough

(Source: Wikipedia [2])

---

2   Wikipedia, Wave. Available ad http://en.wikipedia.org/wiki/Wave

What about the regularity? Firstly, it explains how generations are simultaneously present to the gaze of the researcher investigating them, as they are co-present in social life. The ordinariness of social life itself is essentially given by the coexistence of generations.

The theme of coexistence of different generations has very often been dealt with in pictorial representation. Think of The Three Philosophers by Giorgione (1505-1509) or The Three Ages of Woman by Gustav Klimt (1905). In these works the figures represented together illustrate, also metaphorically, the succession of cohorts, with metaphorical meanings not yet fully clarified. Perhaps it is no coincidence that the first advertising boards dedicated to the promotion of television sets showed television in a household context, with an audience made up of grandparents, parents and grandchildren (The Three Ages of TV Audience)

This coexistence is based on dialectic diversity / regularity: namely that each generation on the one hand recognises the previous or next as different but on the other this difference is attributed to the "naturalness" of the act of generating, of giving birth. What makes this integration possible is precisely the coexistence of different generations within the family and the agencies of socialization. We must therefore assume that both traditional and digital media play a key role. For example, the Catholic Church saw in early television a kind of bulwark in defence of the family, imagining that the nature of consumption within the household would have favoured coexistence and dialogue between children and parents. Before TV the image of the father was of a man going out to a night club, a pub or other premises of "dubious morality". Reading the daily newspaper – the morning prayer of civilised man, as Hegel called it (Habermas, 1991) – was evidently much less familial and intergenerational, but this morning prayer has since become the consumption of television with the proliferation of screens in homes and the different choices of consumption between parents and children. But it could be assumed that currently, with the transformation of TV in the living room into the central terminal, often in the HD version, and thus capable of catalyzing again the attention of the family group, things have changed once more, and generations of parents and children still have some opportunity to live in front of the Tv.

As the diagram of the wave shows, the distance between the crest and the trough changes according to the depth of the water. This reminds us that the conditions in which waves come one after the other may have some sort of effect on their shape. We will return later to this acquisition, but it is worth remembering that media may be not only an intergenerational catalyst but also environmental elements which can constitute the basin where waves follow each other: in other words, precisely those conditions that shape the wave and make it different from others. We can, for instance, imagine that certain stability in the social context tends to produce a different succession of generations from the one produced by a rapidly changing context. Things are even clearer when the regularity of the waves is interrupted by a tidal wave, like a tsunami. It is the situation where a generation not only assumes a peculiar form, but also creates an historical impact, so strong and recognizable as to cause unpredictable effects, designed to have lasting effects over time. This is the best metaphor to describe a generation like that of 1968: a populous generation (son of the postwar period), global i.e. homogeneous worldwide, with characteristic behaviours – the dispute, and most importantly able to portray itself in the unified image through the instantaneous global media of television which has even greater influence than its predecessor, radio (Kurlansky, 2004).

As noted by several authors, media are in this sense actually able to magnify events, as has happened several times, globally and locally, with small and large facts. For example, it is difficult to imagine the social impact of 9/11 without the pervasive power of the old and new media, analogue and digital, which have broadcasted live and worldwide, the collapse of the twin towers. Also, events that are less capable of activating participation and identification may be useful in the construction of generational identity, by acting on the memory of people. This was true, for example, for a global event like the moonlanding in 1969; or facts which in themselves may seem tiny, but which are still able to inculcate among the memories of one or more generations of viewers, so becoming a sort of memorial shared heritage which can guide the perception of time for entire age cohorts e.g. the story of the small child who got stuck in a well near Rome[3]. In these cases, the generations involved in the event as spectators which see themselves as witnesses, different radically from the following generations who only have an indirect account of that event.

This discussion of the metaphor of the wave, then, especially with regard to the exceptional nature of certain generations, gives us the opportunity to see a further aspect of the weight of generations. It concerns both the population abundance that the geographical extension, between local and global. The first element concerns the generation itself, as indeed generated, with no responsibility, but as the effect of the procreative behavior of the previous generation (which generation also takes on the task of caring for children, and thus affecting generationally their orientation). The second element, however, certainly sees the media among its protagonists. They, in fact, develop in the generation the sense of its geographical extension, to allow it to reflect and recognise itself in it. This is an excellent example of how the role of the media can be regarded as firmly intertwined in the construction of generations.

We have seen that the social fact of "generation" can be observed by sociology both in terms of regularity and exceptionality. Each of the two

---

3   The tragedy occured in 1981 kept nearly thirty million viewers in front of TV watching the agony and death of a child (Gamba, 2007).

perspectives clearly points out different facts, though we must say that the role of the media tends to arise particularly when we want to underline the specific and differential nature of a generation among the others. Indeed, the media participate in the dissemination of historical events, and even more, they themselves may constitute, on a smaller scale, events, in that the very appearance of a technology media, of a specific configuration of the media system and so on, can be considered an historical event that, within certain limits, casts a generation, for example by establishing itself as an important shared memory. However, at this point is important to introduce a new distinction in the ways in which the sociological view focuses on generations. Let us take a case history: the birth of independent radio stations, which in Europe took place between the sixties and the seventies, in various forms and with particular histories in each country. That advent was certainly made possible by technological innovation, reduced costs of equipment, some flaws in the legislative and institutional apparatus, all of which gave the opportunity to entrepreneurs, engineers, DJs, musicians and journalists to set up a new 'bottom up' offer for the youth audience. That public therefore found itself in a new position relative to the media landscape, a position that can be described from the newness in the technological, economic and legal scenarios, and the offer and reception of cultural contents. All this emerges from an exogenous perspective, so to speak, from an external observation. But if we ask ourselves how important this revolution was for a generation, we should also ask how newness was experienced, used, told.[4] We could also take a different perspective, endogenous or internal, with startlingly different results. The example I just gave of independent radio broadcasters is interesting because it has points of contact with some of the innovations made possible by digital technology: I am referring to the blogosphere, social networks, the various applications of Wikilogic, where evidently lower costs, exceeded traditional legal constraints and new business models are at work. It may then be interesting to ask why no one has defined the youth of the sixties and seventies as a radio generation, while one can imagine using the term internet generation. One possible answer lies in the distinction between the two perspectives that I just mentioned: the external and the internal.

---

4   An amusing account of this period is given by the movie I love radio rock, (UK 2009, by Richard Curtis).

3 Internal and External Perspectives

We have already hinted at the theme of the external perspective, especially with regard to the description of events that can be considered capable of influencing a generational change. Essentially, we said, they may have different ranges, from the very global (a good example being the fall of the Berlin Wall in 1989) to the strictly local (for example in Italy the abduction and killing of a leading politician as Aldo Moro in 1978), but what counts is the idea that external events can influence and to some extent build a generation. As already mentioned, in this case media play an apparently instrumental task. The diversity of the collective emotional feelings of course also depend on the structure of media systems, because usually we link what we hear to the voice of who recounts it, and to the circumstances of the story. There are differences, therefore, rather significant differences, between observing a media event like the moon landing or the marriage of Charles and Diana, or to learn of an unexpected fact of global nature, such as the assassination of President Kennedy, the fall of the Berlin Wall in 1989 or the 9/11 destruction of the twin towers. But between the last three cases cited, there are also substantial differences in the availability of equipment and the media configuration, given that 9/11 was certainly the first event of historical significance in which the Internet played a key role by speeding up information and its diffusion.

As I mentioned earlier, the aspect of the broadcasting of the event (and that of its construction and co-construction, as in the case of media events) is only one side from which it is possible to observe the generational identity "from outside". Another side, equally interesting, comes from the breeding ground of the generation, as the set of environmental factors, in private and public, that people of the same generation experience together and share.

These factors are obviously both of a structural and cultural nature. For example, the welfare crisis or the labour market reform, with their ability to have more specific implications on some generations than on others (the oldest in the former case, the younger in the latter) are probably crucial in fostering a specific sentiment in different age cohorts. This also means collaborating in constructing a generation by changing the perception of future ones. Institutional and political changes are another of these factors. Beyond the role of the fall of the Berlin Wall, it is obvious that a generation which grew up during the Cold War period sees the world from a very different perspective from someone who did not, and only learns about it (when he does) from school books or through the stories of older people.

The changes and specificities of the social landscape in which people live their youth may also occur at the level of everyday life and culture.

Let us think, in the case of baby boomers, of children's rooms, or to certain types of bicycle, or more generally the existence of certain means of public or private transport, design, and so on. In this sense, media are part of the everyday landscape, both in regards to the public and the private aspects. We might see the different "weight" of different media in different stages, with different interlaces among the media themselves. In the sixties, in Europe, the mix between cinema and television was certainly essential as life experience for children's everyday life. Yet cinemas were very different from today, and their unique form, the rituals in attending them, well described in the film by Giuseppe Tornatore, Nuovo Cinema Paradiso (1988, Italy), are certainly part of their generational identity as a young public at the time. After all, TV was at a primitive stage of technological development, in black and white, with a limited range of programmes and with a rather traditional design which was aimed at incorporating the device into the furniture of the living room. Later, in the eighties, the role of television changed radically, and the new design of the device, colour, the greater availability of channels and programmes, strongly belong to the memory and identity of the young people of those years. Just at that time, cinema underwent a transformation (big spectacular films, the first major innovations in audio systems) and the weighting between the two media changed. Video gaming devices also entered as a new component of everyday's life furnishing. Today, cinema and TV look completely transformed (cinema by multiplexes, by the continuous integration of technology, by a strong resurgence of young audiences; TV by its digitization and by the challenge of new web portals) and the Internet itself has emerged as a new medium, alongside the mobile phone, while the advance of videogames doesn't stop, but explodes with even greater force. It is common concern that the impact of media in households contribute to building a sort of landscape that turns into a perception habit first and thereafter into a shared memory.

I emphasised above the role of rituals, in particular those related to film and television which make up media consumption. But we must not underestimate the rituals related to the media that govern interpersonal communication, from telephone sets and mobile phones to messaging services and social networks. Even in this case we can detect the coding of habits, a sort of familiarity with certain uses of technology. Is it this that marketing points to with its 'deterministic' definitions? Most probably not, for various reasons, of which it is worth recalling the main two.

One is that these rituals (for example, in defining the boundary between public and private, in the rules of courtesy, and in the sense of the value of intellectual property...) depend on a very complex series of elements, in which the presence of a certain technology is just one. Telephone rituals, for example, depend greatly on complex cultural factors, some of which regard more national rather than generational cultures. If in Europe teens use their mobile phones primarily for communication among peers and in Korea for communication with parents, for example, it is clear that the devices do not have a force of their own, but must rather be integrated into the complexity of the cultural identities to which anyone belongs (Thomas et al., 2005).

Secondly, social rituals are encoded as a generational habit when a generation recognises itself as such, and not at the stage of its formation. In other words: young people using the Internet, which build it through certain rituals conditioned by their multiple affiliations (gender, census, territory, cultures and subcultures...), will acknowledge their use of the Internet as specific only in a second phase, when – looking at their past formation – they will begin to look at themselves as a generation. Therefore, the reconstruction of their Internet use will probably be different from what other generations, presently looking at what young do, will perceive as peculiarities. This point is very important because it introduces an additional step in my discourse, about the internal perspective, so important in assessing the generation as a social fact.

The recent sociological reflection on generations insists on the we-sense that is that set of feelings, contents and self-definitions that are shared by members of this collective identity. This we-sense interacts with external definitions given by other social groups (other generations, institutions, marketing companies and the media). It is hard not to recognise, for example, the deeply generational nature of a movie like The Big Chill (USA 1983, by Lawrence Kasdan), which not only tells about an American generation, but which in some way is also targeted to it, through a self-reflexive message. Moreover, this generation will find much appeal in cultural products like the song, and it is worth remembering here an important text like My Generation (1965), by The Who:

> Why do not you all f-fade away (Talkin' 'bout my generation)
> And do not try to dig what we all ss-say (Talkin' 'bout my generation)
> I'm not trying to cause a big ss-sensation (Talkin' 'bout my generation)
> I'm just talkin' 'bout my ggg-generation (Talkin' 'bout my generation)
> This is my generation This is my generation, baby

The ability of a generation to recount itself is, of course, crucial in making possible its sociological observability in terms of an internal perspective. Only through self-narration we can grasp the we-sense. Moreover, we should say that from the sociological point of view, the we-sense exists only in generational self-narration.

In 2008 I published a book called Boom. Storia di quelli che non hanno fatto il '68 (Colombo 2008) which was the result of a series of interviews of my peers (they in fact belong to that generation). The interviews often took the form of real narrations, and I was surprised to notice that even the narrative styles of the various subjects had common points. To present the book, I gave public readings, accompanied by a musician friend and an actor friend. The first performed songs and typical musical themes of our adolescence and youth. The second read excerpts of texts by generational authors about generational events. This caused another wave of self-narrations, confessions, comments, with which I could have written a new book. The reason I tell this tranche de vie is to understand that the ability to recount a generation shows on the one hand how we-sense is fulfilled and on the other how it is is interrelated with the existence of 'public' narrations – like novels, movies, songs and so on – which give the members of the cohort the opportunity to recognise themselves and belong.

We could therefore distinguish between the individual narratives of members of a generation, where you can catch an early form of we-sense, and public narratives that a generation in some way recognises as its own. How to recognise the latter? There are at least two levels of recognition:

The first level concerns the existence of generational successes that is prevalence of consumer products whose audience is eminently recognizable in a cohort of age (Aroldi and Colombo 2007) Where these products are cultural products (movies, books, television dramas, comics, songs, etc.) we believe that they produce a form of reflection which is partial to different degrees, and of self-reflection: in short, they make it possible for people of the same age cohort to pronounce that "we" which constitutes the essential step for the emergence of a generation. It is quite obvious that the culture industry uses instrumentally these reflective opportunities and the charm that this mechanism can exert. Clear proof of this is the specialization of a cultural production for teens, such as the recent chain of Twilights (books, film, merchandising, viral spread, etc.). So are the many youth icons that show-business has built, prearranged to different levels, in its history, from James Dean to Britney Spears (having made the necessary distinctions). However, cultural production – including industrial production – owes its effects not only to the extraordinary apparatus capable of grasping the demand of its audience and achieving their expectations. It also happens that it intercepts almost randomly, instances that are already present, though dormant, which the cultural offer is able to represent.

Beyond these explanations, however, the difficulty of assessing the balance between marketing strategies and authentic cultural self-recognition is the ambiguous generational nature of adolescents of all ages. For they are not yet a complete generation, but only – so to speak – a "promise of generation", which will take form over time. What in fact is lacking in cultural production for teenagers is the voice of the teenagers themselves, who are still at the margin of cultural production: they do not write books, make movies, or speak in radio. They are consumers, but not producers. Not by chance, some scholars, by reconstructing the birth of institutional and scientific focus on adolescence, have spoken about "invention of youth" as an effect of social discourse and organization (Savage 2007).

The reflection of adolescents in the products that cultural production dedicates to them therefore appears generally passive. Quite different is the story of fulfilled generations, which often have ushers who wonder about their affiliation. A good example may be the following excerpt of a very popular song in Italy, written by Giorgio Gaber and Sandro Leporini, and performed by the first. The song is titled Razza in estinzione, written in 2001, and it bitterly reflects on the veterans of the same generation sung by the Who. Here's the excerpt:

> La mia generazione ha visto
> migliaia di ragazzi pronti a tutto
> che stavano cercando
> magari con un po' di presunzione
> di cambiare il mondo
> possiamo raccontarlo ai figli
> senza alcun rimorso
> ma la mia generazione ha perso.[5]

Consolidated generations thus find in media the opportunity to speak through some of their representatives, who can produce a reflection that is not "synthetic" or artificial, but based on common affiliation between the writer and the reader, the producer and the audience. We must say that the development of Web 2.0, social networks and grassroots production, puts us in front of a real turning point. If content production in adolescence was limited until recent times to the private, through diaries, limited socialization, word of mouth, now it becomes visible in the public sphere, thanks to low costs, ease of use of technology and a widespread digital literacy (much higher in younger age cohorts). Here, then, self-expression of young people, the opportunity to share ones feelings and points of view is perhaps the most authentic change in generational affiliation, allowing future generations not to be determined by the technology they use, but rather enabling them precocious we-sense and shared feelings. This tells us nothing about the specific identity of future generations, but reminds us of the integrative role of the media in their construction - a role which, with the changing of the opportunities (for example, offered by the widespread of broadband), does not seem to clearly separate new generations from the old, but rather makes them more alike by accelerating the definition processes of the former.

---

5 My generation has seen/ thousands of kids ready for anything/ they were looking/ maybe with a little presumption/ to change the world/ we can tell children/ without remorse/ But my generation has lost.

Bibliography

Aroldi, P., *Generational Belonging Between Media Audiences and ICT Users*, Infra.

Aroldi, P., Colombo, F., 2007. *Successi culturali e pubblici generazionali*, Milano: Link.

Baricco, A., 2005. *Questa storia*. Roma: Fandango Libri.

Berman, P., 1997. *A Tale of Two Utopias. The Political Journey of the Generation of 1968*, NV: Norton.

Colombo, F., 2003. Does a Web Generation Really Exist?. In: R., Salaverria, C., Sabada, eds., 2003. *Towards New Media Paradigms: Content, Producers, Organisations and Audiences*. Pamplona: Ediciones Eunate.

Colombo, F., 2008. *Boom. Storia di quelli che non hanno fatto il '68*, Milano: Rizzoli.

Craig, S.C., Bennet, S.E.eds., 1997. *After the Boom. The Politics of Generation X*. New York: Rowman&Littlefield.

Di Giulio, P., Rosina, A., 2008. Intergenerational Family Ties and the Diffusion of Cohabitation in Italy. *Demographic Research*, Vol. 16, N.14, pp. 441-468

Edmunds, J., Turner, B., 2002, *Generational Conciousness, Narrative and Politics*. New York: Rowman & Littlefield.

Gamba, M., 2007. *Vermicino. L'Italia nel pozzo*. Milano: Sperling & Kupfer.

Habermas, J., 1991. *The Structural Transormation of the Public Sphere: An Inquiry into a Category of Burgeois Society*. Cambridge: MIT Press.

Kurlansky, M., 2004. *1968: The Year that Rocked the World*. New York: Random House.

Morice, J., 2002. Skills and Preferences: learning from Nintendo generation. In: *Proceedings International Workshop on Advanced Learning Technologies*, IWALT 2000

Palfrey, J., Gasser, V., 2008. *Born Digital. Understanding the First Generation of Digital Natives*. New York: Basic Books.

Savage, J., 2007. *Teenage: The Generation of Youth 1975-1945*. London: Chatto & Windus.

Silverstone, R., Hirsch, R., 1992. *Consuming Technologies: Media and Information in Domestic Spaces*. London: Routledge.

Silverstone, R., 2007. *Media and Morality. On the Rise of the Mediapolis*. Cambridge: Polity Press.

Straurs, W., Howe, N., 1997. *The Fourth Turning. An American Prophecy.* New York: Broadway Books.

Tapscott, D., 1998. Growing Up Digital. The Rise of the Net Generation, New York, McGraw Hill.

Thomas, F., Haddon, L., Gilligan, R. & Heinzmann, P., 2005. *Cultural Factors Shaping the Experience of ICTs: An Exploratory Review*, COST.

Michael Corsten

Media as the "Historical New" for Young Generations

1 Introduction: from Political Viewpoints to Media Systems of Articulation

The 'Problem of Generations' is not recent[1]. 81 years ago, in 1928, a solution was found in Karl Mannheim's corresponding article which has not been surpassed since. Contemporary research into the sociologies of biography, life course, and generations, not least the whole cultural and historical discourse on generation is related to that one essential article.

As the starting point of numerous investigations and debates regarding social, cultural, and historical science Mannheim's (1952) solution to the problem of generations has also often been criticised and considered outdated – but none of these critiques were able to replace the well-differentiated, analytically robust concept of generations as integration of "site", "actuality", and "unity".

However, in recent detailed work on "Historical Generations" by the German researcher Beate Fietze (2008) the observation was made that in the line of studies following Mannheim's approach the originally essential historical dimension of the problem has been lost.

Beate Fietze's observation reveals an empirical problem: we lack 'historical generations' insofar as there are no longer any political generations that share the same generational status, the same historical perspective, and – most importantly – form a unity that would be visible in a particular political movement. The last political movement constituted by a generation was the student movement of 1968[2]. Youths and young adults from the '60s were the last generation in the 'full' Mannheimian sense because *firstly*, they were rooted in a generational status, having been born during the Second World War or in the immediate post-war period; *secondly*, they shared a framework of historical experience: the long lasting '50s and their whiff of narrow-minded Restoration; and *thirdly*, they formed an opposition to the narrow-minded middle-class habits new life-styles which eventually led to new practices in political culture, and thereby renewed our understanding of living in civil society.

---

1 I want to thank Kate Mitchell and Elizabeth Moffat for their efficient proof–reading of my limited English in the draft version of this text.
2 Which besides often is stated to be the first 'global generation'. For general reflections on the concept of 'generation' and on the movement of 1968, see Colombo, infra.

After significant historical events of the '60s took place on a global level, important shifts and transformations were seen in modern world society. The most significant changes to be seen were the beginning of the economic crisis of the mid '70s which was accompanied by an ecological crisis marked by the Chernobyl disaster in 1986; the fall of soviet socialism around 1990 which led to enormous changes in Europe as well as the transformation of Chinese socialism into benevolent socialist capitalism in the same period; and finally 9/11, the attack on the Twin Towers in New York, which became the definitive metaphor for a 'clash of civilizations'.

But despite all these profound historical changes in modern western societies no young generation appeared that possessed all the qualities of the generation from 1968, a generation integrating "status" – social foundation of cohort experience, "actuality" – the framework of a shared historical problem, and "unity" – the formation of shared collective life-styles leading to new political practices.

However, something is missing in this historical description: The technological shifts within media communication since the beginning of the 20th century were not included in the previous description of the historical picture. What of all these radical transformations that followed on from technological shifts in the history of communication media?

Beginning with Mannheim's view of 'historical generations' means accepting a certain idea of political generations. But the meaning of 'political' in 'political generations' might be too narrow. By following this line of thought one tends to assume the idea of 'being political' as types of collective political action, e.g. visible in forms of political movement or protest. The essential expression of these forms of collective political action consists of a gathering of the masses. Gathering of the masses is the leading metaphor of the people's will, of public opinion in democratic societies. But it is – compared to others, e.g. media-mediated forms of publicity – a very romantic type of public expression. It is connected to the bodily presence; it evokes authenticity because all these people are actually in this public space. It is also intertwined with the idea of a struggle: standing up for one's rights, showing up and, if necessary, devoting one's body to the fight for human rights. But, as previously mentioned, such romantic collective gatherings have not been seen in modern Western societies since the 1968 Movement[3].

After consideration of this observation, we may draw two conclusions concerning Mannheim's concept of generations. The first one would be that Mannheim's concept is out-dated because it basically relates to classical pictures of modernity, especially images connecting modernity with liberal and

---

[3] But they could be seen in other parts of the world, only some weeks ago in Iran, for example the killing of Neda as the symbol of the Iranian protests.

democratic political movements. Also, the 1968 Movement would have been the 'last' political movement of modernity, because it had reached the last dimension of liberalization – the democratization of public space and civil society. The second conclusion to be drawn would be that Mannheim's concept of historical generations should be re-written by transforming or ultimately by abandoning the political dimension of generations.

Therefore, it could be useful to analyze precisely if and how the technological transformations in media history had an impact on the formation of historical generations. That is what I will try to achieve here. I want to find the 'historical new' of emerging generations, not in their relation to political development but in relation to media development. In this first attempt to recapitulate Mannheim's theory concerning the relationship between media and generations the question regarding potential political impact is left open. This matter will be addressed in the conclusion.

Firstly I want to reconsider and rearrange Mannheim's concept in three stages. The first is the explication of his main solution which I see as a re-formulation of one basic sociological question: the relation between the individual actor and the social structure, which was conceptualised by Mannheim as a temporal relationship: generational time as the intersection of biographical and historical time. Secondly, I intend to discuss some basic critiques regarding Mannheim's conception concerning the relation between objective and subjective time in his concept and also the relation between meaning and experience in his approach. In the third part I will introduce Foucault's discourse analysis as a bridge between 'subjective' and 'objective' concepts as well as between meaning and experience. Fourthly, and very briefly, I will consider some basic steps and transformations in media history as possible 'challenges' that could affect new formations of generations. Finally, I will conclude by considering the consequences of my thesis, that post-1968 generations are formatted via technological changes, and not by political transformations (in their narrowest sense), according to the social theory of modernity.

2 The Classical Approach to the "Problem of Generations"

As previously mentioned, according to Mannheim, generations show up on three different levels: as (a) 'generation status', (b) 'generation as actuality', and as (c) 'generation units'.

a. When discussing generation status, Mannheim characterises the socio-structural, objective mode of generation, analogous to classes. By using the concept of social status, Mannheim emphasises the embeddedness of generations in socio-historical conditions. In a merely definitional sense, 'generational status' embraces all those people who are born in a certain period of time in a socio-geographically limited space, e.g. people born between 1930 and 1940 in modern Western societies. The sense of the status concept lies in its potential to characterise a structure of opportunities. Two dimensions of opportunities are highlighted by Mannheim: firstly, people born in the same period experience events occurring in the same historical period as well as in the same biographical phase. For example, people who were born in Europe in 1940 experienced the end of the Second World War during early childhood, whereas people who were born around 1930 experienced it during their adolescence. Secondly, people born in the same period experience historical events in a certain sequence during their lifetime. So, for instance, all those born in 1940 experienced the end of the war during their childhood, the reconstruction of society in their youth, economic prosperity in their early adulthood, and a stagnation crisis in their middle age. Mannheim calls this opportunity structure 'Erlebnisschichtung', or 'layer of experience'.

b. It is important to note, that Mannheim subsequently distinguishes between this layer of experience and 'generation as actuality', that is, the way in which the experience of a generation is realised by the generation itself. A member of a generation as actuality displays cognitive or intuitive awareness of the fact that he/she shares a certain structure of experience because he/she has grown up in the same period as other members of the same generation. Individuals of the same age feel associated with each other 'as an actual generation in so far as they participate in the characteristic social and intellectual currents of their society and period, and in so far as they have an active or passive experience of the interactions of forces which made up the new situation.' (Mannheim, 1952, p. 304). Members of a generation as actuality are similar in their perception of what brought about the new situation and also the forces that established the new historical constellation which emerged during their biographical period of adolescence. This specific sense for the constellation and the dynamic forces of 'our time' – the time of a generation – I have called the 'we-sense' of a generation (Corsten, 1999). This second dimension of a generation is the most difficult part of Mannheim's concept. In Mannheim's view it is important that the shared layer of experience given by a generational status does not determine the generation as actuality. As an actuality the members of a generation share a framework of orientation which is led by basic intentions ("Grund-intentionen") and principles of construction ("Gestaltungsprinzipien"). The difficulty in understanding 'generation as actuality' derives from the vague meaning of the original German term "Generationszusammenhang". "Zusammenhang" can be translated as 'connection', 'cohesion', 'correlation' or 'coherence'. Merriam-Webster's dictionary includes an expression of 'coherence' that approaches Mannheim's use of the term "Zusammenhang":

'Coherence = 1: the quality or state of cohering: as a: systematic or logical connection or consistency b: integration of diverse elements, relationships, or values'. "Generationszusammenhang" should be viewed as a state of cohering as integration of diverse elements in at least a threefold meaning: as integration of (1) individuals; (2) groups (units); (3) meaningful experiences. Generationszusammenhang means a shared definition of the historical background of individuals and groups who consider themselves to be part of one generation, and thereby integrate manifold (and meaningful) historical experiences.

c. Mannheim identifies generation units as concrete groups of people of the same age, who not only define their situation in a similar way but also develop similar ways of (re)acting in response to their generational problems. Thereby, an intra-differentiation of one generational context into several possibly hostile or rival generation units emerges. Taking the '60s as an example once again, we can identify 'rockers', 'hippies', 'left activists' and others as concrete groups.

Mannheim reconstructs the emergence of a generation by intersecting two developments, one on the macro-level of society and one on the micro-level. The development on the macro-level is cultural change. The aging and death of the older generations of a society, and the succession of younger generations is interpreted by Mannheim as a vital basis for cultural change. Because the older generation as preservers of tradition and representatives of cultural elites have to stand down and be replaced by the younger generations, Mannheim sees the younger generation as a social group which is able to accept responsibility for the cultural storage of a society in a new way. The young generation is the one social group which is able to re-interpret the cultural repertoire of a society in its most recent state. They are able to relate to culture in its current (most developed) state of the art. On the micro-level the young generation resides in its biographical state of youth, or as we say today, in adolescence. In this biographical phase, young people are highly aware of becoming adults, being responsible for themselves and for finding their own way in life. In this phase Mannheim supposes the generation – as an expression of a collectively shared life course – is sensitive, or as he puts it: impressive – regarding the historical new.

Mannheim concludes that the formation of a generation, therefore, is a match of developments in time. The cultural development of a society which historically produces new forms of style matches the collectively shared biographical time-period of adolescents who are sensitive or impressive towards historically emergent forms.

2 Basic Critiques/Questions on Mannheim's Concept

By emphasizing the time issue of Mannheim's concept we are able to reject some critiques of his approach but we are also confronted with new difficulties.

Firstly, we can argue that critiques of Mannheim's similarity to Marx's concept of class society, especially to the problem of class consciousness can be rejected. Generations – unlike class – are not constituted by inequalities of social resources, or particular group interests. Generations are constituted by their position in social, cultural, and historical time. Therefore, they are not necessarily political actors. For Mannheim, generations are basically defined by their relation to (or placement in) the cultural process of modern society.

Taking the example of the generation from 1968, for Mannheim, it would not have been crucial for this Movement to become political. Mannheim would have looked at the cultural roots of this Movement. Consequently, the constellation of a generation entering adolescence at the end of the fifties had to be described according to its relationship with the emergent forms of culture in the sixties. Then the coherence of the generation from 1968 is constituted by the radical break that they made with the past (the Fifties); and the radical changes they made by adopting trends from the Sixties. It is this tendency towards radical change in which we find the leading principle of generational self-configuration of the 1968 Movement; or in Bob Dylan's famous song (born 1941): "your old road is rapidly agin' – please get out of the new one if can't lend your hand – cause the times they are a-changing'" (1964).

The 1968 Movement had transformed their shared historical and biographical experience into a new style of self-expression. Less important than the political impact of this collective self-expression was its constitutional dependence on the communication media of popular music culture.

But if the emphasis lies on time then what has to be asked is which concept of time is assumed in Mannheim's approach on generations? Is it an objective concept of time sequences, as in a series of life events or a succession of life phases? Or is it a subjective concept of time, as one of biographical perspectives? Or are both dimensions of time integrated – a chronical of states and events and the inner perspective of experienced time – and if so, should it be interpreted as a synchronous or as an asynchronous intersection between the order of events and the formation of time perspectives?

Once more these questions reflect the relations between generation as 'site', or layer of experiences and generation as 'actuality', or the configuration of a shared generational background. The generational status could be interpreted as an objective order or sequence of experienced life events and biographical phases. But a generation as actuality includes the moment of self-reference. As an actuality, generations are able to attribute the experienced time to be 'the time of our generation', or as 'our time'. The difference is that the experience of the members of a generation does not only take place in the same manner. The members of a generation share a 'we-sense' by being able to rediscover the individual experience of biographical time in the ways in which other members of his/her generation express their biographical time, and then transform this shared perspective into the time of 'my generation'.

But then it has to be explained how the factual flow of experience could be transformed into a configuration of experience shared by the members of a generation. Obviously, the knowledge that other members of my generation share with me, the same layer of experience can not be directly transmitted via experience. A kind of communication is needed to explain how a certain configuration of experience becomes meaningful to us as members of the same generation.

Initially, this does not seem possible within Mannheim's framework of analysis. Prominent critiques by C. Wright Mills, Tim Dant, or Jane Pilcher have argued that it would impossible to reconstruct the symbolic dimension of this problem in the framework of a sociology of knowledge.

While it is true that Mannheim lacks an elaborated communication or discourse theory, it would also be impossible to combine Mannheim's sociology of knowledge directly with a discourse theory. This is because Mannheim himself was highly aware of some of the problems concerning sole communication analysis, especially in his early writings on culture and on the 'structures of thinking'. From Mannheim's point of view not only the processes of thinking or the formation of knowledge have to be considered as socially, culturally, and historically constituted. The formation of language and communication in particular has to be analyzed as an historical and social circumstance.

As for the problem of how generational experience can emerge, it cannot be solved by saying that members of a generation are able to attribute their experience as a collective one just by using symbols, signs, language, communication, discourse and so on. Mannheim's argument here consists in a deeper insight, that using signs, acting communicatively or in discourse practices are not universally structured regulative mechanisms (as is supposed by Chomsky in the analysis of transformation grammar or in Searle's approach to speech act theory). Using signs and language and communicating is rooted in historical preconditions and relatively constituted by the members of a society in which signs, language, and communication emerge in a historically specific way.

In his early "Essays on the sociology of culture" Mannheim introduced three concepts which bridged the gaps between experience, knowledge structures, and discourse formation: "Kulturgebilde" (figures of culture/artefacts); "konjunktive Erfahrung" (conjunctive experience), and "Bedeutsamkeitszusammenhang" (coherence/inner context/network of meaning).

In Mannheim's early approach, signs, language and communicative actions would be 'figures of culture' that were associated with an 'inner context' or coherence of meaning. Basically, figures of culture were carried by processes of conjunctive experience. Within the concept of conjunctive experience Mannheim tries to explicate that human experience, thinking, and knowledge is co-constituted from the beginning by shared experience and shared meaning which is also body-related and world- (materiality-/object-) related. Mannheim's concept (of the intersection of body-based experience that could be shared conjunctively and is co-constituted by meaning attributed collectively to a social and historical situation of cognition) is not dissimilar from more recent approaches regarding post-structuralistic discourse theory. As Foucault (1969) describes the formation of discourse as the intersection of materiality, articulation modes, strategies, and concepts, Mannheim also emphasises an intersectional view on the meaning-network of cultural figures as bodily-related, socially experienced and meaning-attributed world features.

As a result, Mannheim's sociology of knowledge is not a cognitivistic approach. The dimensions of experience, thinking, and knowledge are always constituted by a (biophysically based) body-world relation from which experiences derive and which are fundamentally socially shared and meaning-attributed. These three constitutional dimensions: body-world relation, social conjuncture of experiences and its association with meaning cannot be divided. Experience is formed by the intersection of these three features.

In a very basic sense one could also say that these three features: body-world relation, conjunctive experience of a social community, and the meaningfulness of experiences are fundamental media of experience. Or to use a different formulation: experience is not an incidental case occurring in the world instantaneously; it is a mediated relation between perceptive organisms, social organisations, meaningful artefacts and the biophysical world.

Taking this basic understanding of experience and culture into account, media as "cultural figures" or "artefacts" are always relevant, not least in regard to the problem of generation.

3 Introducing Media (Luhmann) and Discourse Analysis (Foucault)

In the third part the relationship of communication media and the discursive formation of generations will be explored and clarified.

First of all "media" can be defined as (mostly technical) ways of distributing or disseminating communication. Media should also be regarded as artefacts recording, and thereby documenting communication and other forms of shared symbolic practice. Media are means of publication – modes to include an audience into the signing or symbolization of world features – into a process of 'conjunctive experience'. Modern communication media are means of mass media which afford world-wide communication. Mass media communication generalises and standardises the production of meaning and individualises the perception of communication. Media enable one to copy, to duplicate, to multiplicate culture figures – e.g. the original 'la cena' of Da Vinci and all of its copies.

A very important feature of modern mass media is also its capacity to record and thereby document signs, symbol practice, and communication. Social experience and the subsequent societal usage of signs have become historically reversible. Just by using the replay button images of past events and symbolic actions can be repeated again and again.

The possibility of scripture has offered society high reflexivity and precision in the analysis of symbolic practices and cultural artefacts. Nowadays, together with the support of photographic, audiographic, and videographic recording techniques, the analysis of symbolic practices and cultural artefacts has become a symbolic practice in itself with its own 'logic'. Not only does this result in a kind of world duplication, media also enable us to perceive and experience 'things' which otherwise would not be visible or audible without them, for example slow motion and soccer.

Taking these general capacities of mass communication media into account one could go on to relate the historical shifts in mass media in the 20th century to Foucault's analysis of discourse formation. Let us not forget that media transform the ways in which the experience of the world can be socially and symbolically mediated. They increase the amount of world features which can be experienced; they increase the amount of people with whom we can share experience; and they increase the ways in which experiences could be symbolically organised into a framework (configuration). In this way media shifts offer new events in discourse.

As previously mentioned, Foucault analyses discourse as an intersection of four levels of discourse formation: formation of materiality; formation of modes of articulation, formation of strategies and formation of concepts.

With his formation of materiality Foucault emphasises that the discourse is not only effected or caused by the material (biophysical) qualities of the world but the discourse also structures the material world by regulative mechanisms of discourse. By the circumstance of articulation modes he points out that the expressions of actors are not primarily reflexes of their inner states and feelings but that the discourse offers modes to symbolise or articulate the meaning of the 'invisible souls.' When discussing the formation of strategies Foucault contradicts the concept of psychic motivation (mind), that people are driven by universal needs. Alternatively, he tries to show that the needs, interests of actors depending on their social positions, and the strategies which are associated with them., Finally, the knowledge system should not be regarded as a consistent conceptual scheme derived from a pure operation of thinking ('reine Gedankenarbeit'). Concepts are results of their placement in an historical and social context of other concepts, and they are over-determined by techniques of proof, by social strategies, or restricted to the modes of articulation allowed in the language game played by society.

4 Changes in the Formation of Media Discourse

Turning now to the analysis of media shifts during the 21st century I want to follow the line of the popular music genre as a stable mode of articulation which is also important for the young generations of a particular historical time-period. From this level I want to look at how changes in the materiality of discourse (recording techniques, replay techniques, broadcasting techniques) are intersected with changes in discourse strategies, and concepts.

My argument will be that future young generations play the role of elaborating the symbolic possibility of new formative capacities in discourse formation. Therefore, they are curious about new technical possibilities, and they use them to establish new types of articulation in discourse. These attempts at articulation are accompanied by strategies of cultural competition, and by the composition/configuration of new cultural concepts.

For the sake of brevity, my historical thesis is illustrated by two diagrammes/tables.

The first diagram illustrates the main shifts in communication media. The focus on these shifts relates to the questions: How is an expanding (worldwide) audience informed? What are the basic media for distributing popular music?

Time	Materiality	Articulation Mode	Strategy	Concepts
1920-50	Radio Days	Studio vs. Live	Reproduction	Purity
1950-70	TV, Records	Electrification	Pre-domination	Intensification
1970-80	Taping the World	High Fidelity	Distinction	(Im)Perfection
1990	WWW 1.0	Remix	Integration	Harmony
21st Century	WWW 2.0	Appearance	Networking	Performance

Table 1: Shifts in Communication Media (relating to popular music). (Source: the author)

Five basic changes should be noted. The first shift is the establishment of a radio broadcasting culture in the 1920s. The basic strategy is reproduction. The main difference is between live or studio music production. Also, the first examples of very fragile records (shellacs) were produced and distributed. The concept integrating these modes of articulation is purity.

After the Second World War, between the 1950s and 1970s, television broadcasting and records (singles) dominated. The so-called 'juke-box' was a very important medium. In many countries watching television and listening to records took place in public spaces, watching TV in pubs, listening to the jukebox in milk bars or ice-cream parlours. The electrification of music instruments – mainly guitar and bass – was the dominating articulation mode in popular music. Tube amplifiers such as the famous Fender Super Reverb supported a strategy of 'pre-domination', of drowning out surrounding sound, and of course of amplifying. The concept was the Intensification of Sound.

Since the mid-70s, and especially during the 1980s, forms of taping were increasingly elaborated. The portable radio-recorder containing a micro-cassette became famous. The possibility of recording music from radio introduced actors to a more individualised or privatised consumption of music. Individual mixes of music pieces became dominant and a source of distinction via music and/or life style. The elaborated modes of taping, e.g. in high fidelity sound, emphasised the conceptual difference between seeking (artificial) perfection or allowing (authentic) imperfection. With the emergence of computers, compact discs, MP3 players, music became more and more remixable. DJs worked on two or more turn-tables to patch and scratch pieces of tracks together to form endless pieces of music. This strategy of patch-work integration led to a harmonic concept of music which aimed to include all the differences of a culturally-diversified, popular music scene. Music is now available all over the worldwideweb.

In my opinion, it is not clear, if one can really talk about a difference between World Wide Web 1.0 and World Wide Web 2.0 in a convincing technological sense. But what we can empirically observe is the usage of combined multiple media of communication: computer, cell phone, internet, dvd, mp3-player – and sometimes also the good old fixed-network phone (see Table 2). This multitude of options might have led to more unobtrusive strategies: neither integration nor even harmony are objectives. It is deemed sufficient to articulate oneself by simply appearing somehow, sometime, somewhere for somebody and to perform in a suitable way. Showing up sometime, someplace for somebody also satisfies the strategy of networking in an increasingly complex world.

	1963	1969	1973	1978	1983	1988	1993	1998	2003	2008
TV	34,4	72,7	87,2	93,2	93,8	94,9	95,6	95,8	94,4	94,1
Narrow-film camera	1,8	4,8	8,4	12,7	13,2	10,5				
Photocamera	41,7	61,0	68,4	74,9	77,8	76,7	79,3	86,3	83,4	85,2
Radio-recorder	5,1	19,1	25,4	18,9	34,9	37,1	9,6			
DVD-Player									27,1	69,1
MP3-Player										37,3
Telephone	13,7	31,0	51,0	69,5	88,1	93,2	87,3	96,0	98,7	99,0
Computer							21,2	41,5	61,4	75,4
Internet								8,1	46,0	64,4
Handy								11,2	72,5	86,3

Table 2: media possession in households (in % of households in Germany).
(Source: Household Statistics, Statistisches Bundesamt)

5 Conclusion

Leaving the shifts in media usage and returning to the problem of how generations are constituted presents some concrete answers:

Families constitute generations as long as we distinguish parents and children. Politics and culture might constitute generations as long as a distinction is made between old and new, tradition and progress, ...

But do media constitute generations? And, if so, how?

In line with Mannheim's view on generations as site, as actuality and as units, generations are not constituted by the cycle of families or by a linear process of modernization. Generations constitute themselves by defining the historical new situation via re-configuration. The shared cultural traditions of a society are re-interpreted and re-configured by the youngest generation. Therefore, media developments offer new opportunities to mediate collective experience.

This is the basic argument of my reconstruction. The 'Historical New' of a generation is not a substance, a thing in itself. The 'Historical New' is a perspective, a 'practical sense' (Bourdieu, 1980) for the new constellation shared by the members of the youngest generation. The shared awareness for the 'Historical New' is also the 'we-sense' of a generation with which they identify themselves by facing the meaning of 'our time'.

But the meaning of 'our time', the awareness of the 'Historical New' have to be mediated. Therefore media shifts offer new materialities, new modes of articulation, new strategies, and new concepts. All in all these media shifts offer opportunities to re-arrange, to re-configure and to re-interpret the cultural repertoire of a society as an on-going process.

Bibliography

Bourdieu, P., 1980. *Le sens pratique*. Paris: Ed. de Minuit.

Colombo, F., *The Long Wave of Generations, Infra*.

Corsten, M., 1999. The Times of Generations. In: *Time & Society*, 8 (2), 249-272.

Foucault, M., 1969. *L'archéologie du savoir*. Paris: Gallimard.

Fietze, B., 2008. *Historische Generationen*. Bielefeld: transcript.

Luhmann, N., 2000. *Reality of Mass Media*. Stanford University Press

Mannheim, K., 1952. The Problem of Generations. In: Mannheim, Karl (edited by Paul Kecskemeti): *Essays on the Sociology of Knowledge*. London: Routledge and Kegan Paul pp. 276-320.

Mannheim, K., 1982. *Structures of Thinking*. London: Routledge and Kegan Paul.

Piermarco Aroldi

Generational Belonging Between Media Audiences and ICT Users

1 Talkin' 'bout Generations (and Media)

This article aims to present some ideas about the relation between "media" and "generations" and, in a more specific way, between ICTs and generations of their users (or non-users). The main questions I will attempt to answer are: firstly, whether (and how) media take part in the construction of generational identities; secondly, whether (and how) generational belonging affects media usage and sense-making in the everyday practices of media consumption; finally, what the differences introduced by ICTs in this relation between "media" and "generations" are.

The underlying assumption of this article is, obviously, that talking about "generational identity" or "generational belonging" is made possible by the real consistency of social formations called "generations": by this word I mean, in agreement with Edmunds and Turner, "an age cohort that comes to have social significance by virtue of constituting itself as a cultural identity"(Edmunds & Turner, 2002).

Notwithstanding this assumption, I am less interested, here, in discussing the *actual* generation *labels* and denominations (like *Boomers*, *X Gens*, *Generation Me*, *Millennials* and so on) or the *peer personality* of each of them (as described, for instance, by Smith and Clurman (1997), Strauss and Howe (1991, 2000) or J.M.Twenge (2006)) or the *time span* dividing one generation from another , than the general dynamics of the generation-building processes and the *roles* played in them by the media: my point is, in fact, that the media play different roles at different moments of this social construction of a shared identity, and that these roles are strongly affected by a lot of variables, both socio-cultural and technological; so that talking about "Net generation", for example, may be a lucky but really misguiding keyword to address youth, even in the western world, because of its oversimplification.

In order to reach these goals, I am going to propose a model of the generation-building process based on a definition of "generation" which follows that found in a review of some of the scientific literature and in a research project entitled "Media and Generations" which was funded by Italian Miur (Ministry of education, university and research). I will then describe the roles played by the media in this model and in different theoretical approaches. Finally, I will discuss the new developments introduced by ICTs.

1.1 Generation's Coming Back

To begin with, there is a notable return of the topic "generation" in the international debate and research of the last ten or fifteen years; the reasons for the interest regarding this issue are probably various: on the one hand, there is a crisis of the traditional forms of collective identity and social belonging, such as class or national community, really put to the test by the change of our late modernity, and the resulting need for researchers to identify new categories that could help us to understand the social phenomena (Dandaneau, 2001) or to forecast consumers' behaviours (Smith & Clurman, 1997). On the other hand, the speedy development of ICTs and media seems to have radically changed the forms of cultural transmission and socialisation, stressing gaps and differences between social groups and between age cohorts. One could say that a sort of "digital generational" rhetoric inspired a plurality of discourses, from marketing to education, to highlight the need to find new ways to address new consumer groups (Smith & Clurman, 1997) or new classes of students (Prensky, 2001). Some of these discourses will be discussed in the following pages.

At the same time, sociologists and historians seems to have rediscovered such a "traditional" category too: the former returned to classical sociology, from the seminal Mannheim's essay (1952) to the Bourdieu's research into professional-artistic or academic fields (1988), trying to read the new dynamics of social change in the key of generational consciousness (Pilcher, 1997; Edmunds & Turner, 2002); the latter reacted to "The End of History" trying to "apply the mirror of recurring human experience to gaze around the corner of current trends and say something instructive about the decades to come" (Strauss & Howe, 1991) and claiming "how today's small children lie not at the end, but near the *beginning* of a new generational cycle" (Strauss & Howe, 1991). So, describing the new Millennials, Strauss & Howe put themselves in the same line as Elder's work on the generation that grew up during the Great Depression, a traumatic event (Alexander, et al., 2004) that united a particular cohort of individuals in a conscious social stratification based on age (Elder, 1974); or of Wyatt's studies of the American generation of the Sixties, interpreted as a social group particularly active in determining, through the affirmation of a true generational subculture, an unexpected acceleration in the American society (Wyatt, 1993): a theoretical background soon easily emphasised by "9/11" attack to WTC.

Against this background, media studies tradition *before* ICTs studies seems to focus very little on the topic of "generation": at a glance, it is possible to propose a generational interpretation of Meyrowitz's theory about Television and electronic media as drivers for social change in the USA (Meyrowitz, 1985); there are some traces of generational belonging as a variable taken into account, besides class and gender, to better comprehend the Television experience of the women in the audience studies tradition (Press, 1991; Heide, 1995); there are a lot of cultural histories of the early ages of radio and Television on the basis of private memories of the first generations of listeners (Moores, 1988) and viewers (Kortti & Mähönen, 2009; O'Sullivan, 1991).

1.2 A Map to Generations

OssCom's research on generational belonging in media audiences started in 2001, when our working group tried to find out the peer personalities of four spectatorships' generations in Italian Television; then we focused on the notion of "generation" itself, trying to understand the social making of the age cohorts as cultural identities implying media uses and media contents as tools and materials of this building; finally we returned once again to the field questioning different generations of people about their formative memories of viewing / listening /playing / using media and ICTs to find out their presence in what Corsten calls "generational semantics" (Corsten, 1999).

Both the roles played by the generational belongings in determining media diets or media sense-makings and the roles played by the media in defining generational identities and "we-sense" (Corsten, 1999) were investigated.

During these years our conceptualisation of what a generation actually is has changed, moving from a marketing oriented definition to a more sociological one: a multi-dimensional concept, where biographical traits shall coexist alongside historical, biographical and cultural ones, and where age group belonging is connected to specific historical experiences, to the development of peculiar consumption habits or to the occupation of certain positions in the family chain. Such a multi-dimensional category appears particularly useful within a theoretical and research paradigm, in which the different segmentations of the consumer body (audiences in this case) cannot be reduced to either individual socio-demographic traits (such as age, gender, education, job position) or the corresponding life styles (such as those codified by marketing), but have to be strictly and simultaneously related to several factors - such as one's position along the lifecycle, media biography, contexts provided by families and friendship networks as environments for the elaboration of media experience, the belonging to a world of values shared with other members of the same generation, the historical development of the media system, the different phases of technological innovation, the processes of taming and incorporation of technologies and media products, as well as the wider structural changes affecting the social and cultural system.

Karl Mannheim's elaboration (1952) obviously appears useful in highlighting many important points. It is common knowledge that Mannheim's proposal distinguishes between "generation status", "generation as actuality", and "generation unit". The "generation status", like class status, is a simple bond among individuals who, being born in the same period, occupy the same social space; thus the generation differentiates itself from the formation of concrete groups based on either communitarian structures (family, tribe, lineage) or associational ones (groups built around an intentionally subscribed goal, law, or rule). Each generational collocation in the social space can be associated with an inclination towards behaviours, feelings, action and thought models. The "generation as actuality" represents something more than a collocation: it is a further bond: "this additional nexus may be described as participation in the common destiny of this historical and social unit". A "generation as actuality" should be thought of as an actualization of the potential implicated in the simple "generation status". The very "generation units" (which can take the form of concrete groups) are grounded upon the sharing of this bond.

Affinity among the individuals belonging to the same "generation unit" reveals itself in the sharing of the common contents forming individual consciences and acting as in-group socializing factors: in a word, the Gestalt – the peculiar way of perceiving, interpreting and evaluating social, historical, and cultural phenomena.

Or, in Mannheim's own words, "youth experiencing the same concrete historical problems may be said to be part of the same actual generation; while those groups within the same actual generation which work up the material of their common experiences in different specific ways, constitute separate generation units" (Mannheim, 1952). Belonging to a "generation unit" means sharing a particular "unified view": "it involves the ability to see things from its particular 'aspect', to endow concepts with its particular shade of meaning and to experience psychological and intellectual impulses in the configuration characteristic of the group ... to absorb those interpretive formative principles *which enable the individual to deal with new impressions and events in a fashion broadly predetermined by the group."* (Mannheim, 1952)

Michael Corsten builds his commentary on Mannheim around the focalization on the feeling of generational belonging: what is it based upon? How does it develop and sustain itself? To find an answer to these questions he refers to the concept of semantic history, i.e. the idea that a generation recognises itself as such when it is able to produce a dominant order of meanings continuously empowered through discourse practices and significant rituals among the members of the generation itself. Generational semantics is, in other words, a collection of themes, interpretative models, evaluation principles and linguistic devices through which shared experience is transformed in discourse within the forms of daily interaction. It can be read as a process of crystallization of a generation's encyclopaedia and the linguistic rules employed to consult it, as well as a social contraction: the "cultural circle" of a generation is not a real, concrete group of individuals; it is not a "generation unit" but a bond, hence, in Mannheim's terms, a finite number of coetaneous individuals who "spontaneously observe that other people use certain criteria for interpreting and articulating topics in a similar manner to themselves" (Corsten, 1999). Corsten returns to the topic of Mannheim's age of adolescence, expanding the nature -not only biological but also historical and social- of this stage of the life-cycle, in which the issue of personal identity is felt with particular urgency and prominence. Experiences during adolescence are important in defining the we-sense of a generation, not so much because the age is in itself more "sensitive" or "impressionable", but because the common historical, social and cultural context provides a wide group of adolescents with the same tools to define their own individual self – starting from the very generational semantic that makes up the common language as well as the thematic repertoire to reflect one's forming identity with.

1.3 A Model for Generations

In light of this sociological literature on the topic of the generations, the main outcomes of the "Media and Generations" project allow me to propose a very schematic model aimed to describe, from a phenomenological point of view, the complex processes of generational identity-making and the roles played within it by the media.

This draft model cuts off, "freezing them" in a very artificial way, the "fields strength", operating in the past and in the present to shape the generational identities, thought as an open "work in progress" but characterised by some - more or less- stable and shared features. To this generational identity belong values, ideals, configurations of taste and sensibility, constellations of preferences that we could probably call, with Bourdieu (1979), *habitus*, that is a system of durable dispositions to act and to choose, not strictly prescribed by formal rules, for example in the field of civic participation, of material and cultural consumption, of leisure.

In such a model I have tried to represent, first of all, the two "fields strength" (on the left and on the right) characterised by an exogenous dimension: an objective status, caused by the fact of being born in a certain historical moment.

The field on the left is developed in the past, during the formative years, and includes historical and political events, material and symbolic constraints and resources differently distributed among the population, some of which commonly shared by all -or the majority- of its members, others reserved to a minority or to the elites on the basis of the social stratification or other socio-demographic variables like age and gender. To be considered significant, in this area, on the one hand, traumatic or catastrophic events, able to mark a high discontinuity between "before" and "after", or a particularly intense experience, good or bad, within individual biographies (wars, revolutions, crisis etc.) and, on the other hand, cultural institutions and processes, commodities and services that shape the everyday experience, especially if intended for a young "audience": educational systems, formative agencies, social practices, technologies, brand, media and their contents.

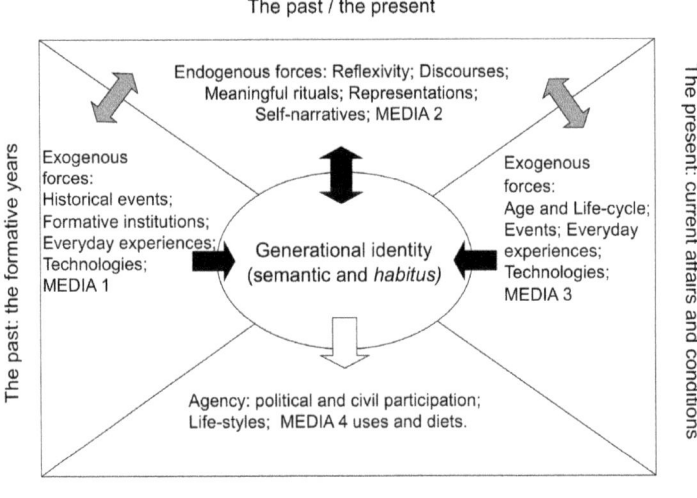

Figure 1: A model of generational identity-making. (Source: the author)

The field on the right, exogenous as well because tied to the generation status, includes historical events and contingencies, social and cultural constraints and resources given in the present (obviously, in the "contemporary present" of the moment in which the model is applied). First of all, age and position in the life-cycle, combined with the biographical conditions depending on such a generational location and, for this reason, shared with the other members of the same generational cohort. People who had the same experiences in their formative years have thus, nowadays, to face with historical events and social conditions that are common to everyone who lives in the same society; but that people filter, and make sense of, these present experiences in a very peculiar way, shared within the same age group only, and depending on both a social position and a cultural identity: their generational identity and semantic, their we-sense.

Following Mannheim (1952), Corsten (1999) and Eyerman and Turner (1998) a third "field strength" must be taken into account (on the top, overlapping with the first two of them), acting both in the past and in the present and characterised by a more "endogenous" perspective: it is the field of reflexivity, of self-consciousness, of mutual identification by the members of a generation. In Mannheim's analogy with class status, in fact, generation consciousness stands like class consciousness, and it is not necessary to belong to a "generation status" as a social collocation: "inherent to a *positive* sense in every location is a tendency pointing towards certain definite modes of behaviour, feeling, and thought" (Mannheim, 1952). But this tendency lies in a "stratified consciousness" (Mannheim, 1952), where "early impressions tend to coalesce into a *natural view* of the world. All later experiences then tend to receive their meaning from this original set, whether they appear as that set's verification and fulfilment or as its negation and antithesis" (Mannheim,1952). If it is true that this "natural view" acts like a "common sense", by definition a "non-reflexive" set of knowledge, on the other hand, Corsten introduces the dimension of reflexivity because the members of a generation share not only a common background of experiences, but above all they share the consciousness that also the other members of the same generation share the same background. As Corsten says, "they do not only have something in common, they have also a (common) sense" - in the twofold sense of *shared* and *taken for granted*- "for the fact that they have something in common." (Corsten, 1999). This reflexive "we-sense" contributes to the making of a generation because a generation is formed exactly "by the sense that the members of a generation have of the criteria for belonging" (Corsten, 1999).

Where does such a form of collective self-identification actually take place? As previously stated, a "generation builds up its comparable interpretations by establishing a dominant semantic order (order of meanings) in its discursive practices" (Corsten, 1999) maintained, both during its adolescence and the following years, through rituals, narratives and mythologies, and fed with self-representations as well as with representations made by the other generations and with the other generation's representations: self identity, in fact, is always negotiated with (and against) others' identity and their mutual representations. These reflexive and discursive practices use, on the one hand, the materials provided by the past, shared, experiences and, on the other hand, the hints and the inspirations provided by the present occasions. Generational identity results thus, in this model, as a social construction rising from both some exogenous forces and some discursive practices, mutually interwoven; a social construction able to affect cultural belongings and to shape a peculiar *habitus*, thought as a "collection of practices through which generational experiences are manifest." (Edmunds & Turner, 2002)

Finally, the last area (at the bottom) of the proposed model represents the sphere of the social agency, where political issues are managed and acted out, lifestyles are followed, brands and commodities are preferred showing, in a very evident way, generational identities and belongings. Stronger is this generational identity, and wider – and deeper – is the sharing of its *habitus*, a certain set of choices more likely depends on a generational belonging than on simple socio-demographic attributes.

1.4 What about the Media?

General hypothesis of our "Media and Generations" project was that the media play a role in the processes I have attempted to describe; the main outcome of our research is that the media play different roles in each of the "field strength" represented in the model. First of all, media (as technologies as well as contents, as in Silverstone's concept of "double articulation" (Silverstone, et al., 1992; Silverstone, 1994)) have a part in defining the formative experiences of a generation, not only because they are so deeply embedded in the everyday practices as to become a "natural" element of its social landscape and its common sense, but also because historical events and occurrences, as well as cultural values and their symbolic forms, are often mediated by them. Media ("Media 1" in the model) provide a lot of the material and the tools used in the making of the generational semantics: news, facts, imageries, characters, celebrities, emotions, rituals, icons, music and brands that lie in the memory in a very stratified way and that contribute to shape (and to share) cognitive patterns, tastes, attitudes and expectations.

Obviously, all these elements act in the past of the formative years and survive in the memory just as fragments, echoes, and scraps ready to re-emerge if "stimulated" by the present circumstances (or by a social researcher).

When they do re-emerge, they often bring with them a part of biography (which they are linked to, as in the case of the "flashbulb memories") and a set of previous experiences to compare with the others' experiences. During our qualitative research, based on life histories and focus groups, our respondents revealed mnemonic *repertoires* filled with a lot of media-related materials (titles of TV programmes and movies, names of musicians and rock bands, comics, cartoons, sport events, catchphrases, jingles and so on) very useful for them to tell their personal history as well as to recognise, mutually, people belonging to the same generation.

In the "present" area of the model proposed, the media do act in a different way: not as "natural" providers of "building material" but as "catalysts" and "hooks". Explicitly, media ("Media 3", in the model) are a part of the present experiences (conditioned by the social positions in terms of age and lifecycle) that interact with the stratified memories and are coped with by them; sometimes the new technologies and the new contents react with the generational semantics and identities producing new forms of feeling (as in revivals); sometimes they force memories to re-emerge; sometimes they simply cut off (or catch in) a target group.

However, it is probably in the "field strength" of reflexivity that the media play a very different role; in this area they are resources articulating the public spaces of the generational discourses. Here the media ("Media 2", in the model) act like available technologies and institutions that make it possible for the members of a generation to develop its discursive practices, its mythology, and its representations. They connect people with each other, giving mutual visibility, allowing people to tell their story to each other and assigning leaderships and entitlements to "speak on behalf of" the other members. They contribute to establish (and sometimes they simply *are*) the meaningful rituals where generational contents are shared and interpreted, appropriated or contested; they are the stage where the mutual generational representations are acted out. According to Corsten, the media seem to be (some of) the main discursive resources that make the members of a generation conscious "that they have something in common" (Corsten, 1999), and that this "something" is really different from what is "common" for their parents (or children) and for their older (or younger) brothers and sisters. This reflexivity works both on the materials of the formative years, during the adolescence and youth ("Media 1"), and on the new materials of the nowadays ("Media 3"), embedding or rejecting social meanings and values to build a very dynamic "generational semantic".

Finally, "Media 3" are, as well, the providers of a set of proposals amongst which people make their choices: on the basis of their *habitus* and conditioned by their present social position, members of a generation choose some media and reject some others, *domesticate* in a certain way a new technology, set up a determinate "media diet", use (or do not use) a kind of device, read and interpret some narratives. "Media 4" represent thus the media practices actually performed by the members of a generation; forecasting these behaviours and providing them with generational products and rhetoric, are, obviously, the main aims of the generational marketing.

2 Generations, Media and the Novelty of ICTs

The wide diffusion of ICTs, in the last two decades, coincides with the rise of the so-called "Internet Generation"; many popularizing commentators introduced generational definitions based on the similarity between people's childhood and the metaphorical childhood of the emerging technologies. "Web generation", "digital generation", "bit generation", "Nintendo generation", "e-generation" and so forth are just a few among the many variations on this theme, based on the assumption that the younger generation are supposed to be more naturally inclined to get accustomed to new technologies than the elders.

The idea of an "Internet Generation" (or, quoting Hartmann (2003), the "web generation discourse"), seems to have had, in these years, two main fields of application; the first one is the marketing discourse, pushed by the need to find, show (and sell) the newest target segmentation. The second one is the educational discourse, curiously often very close to the conceptualization offered by the first one in its facing the fast transformation that affected skills and competences of the "new" learners. Here I can just remind some of the authors who developed this idea: Tapscott, on the business strategy front, for instance, proposing his "Net Generation" for the Echoboomers, states that their defining characteristic was that they were the first to be "growing up digital" and they were different from any other generation because they were the first to grow up surrounded by digital media: "Today's kids are so bathed in bits that they think it is all part of the natural landscape" (Tapscott, 1998), and this condition makes them a force for social transformation; ICTs, in fact, are seen to create new styles of communication and interaction, new conditions of education and learning, new competences and skills, new forms of personhood and new tendencies in political participation. On the educational side, Prensky calls "Digital Native" today's students, grown up with ICTs, "native speakers" of the digital language: they "are used to receiving information really fast. They like to parallel process and multi-task...They prefer random access. They function best when networked. They thrive on instant gratification and frequent reward" (Prensky, 2001). According to Prensky, their thinking patterns have changed because the digital input they received when growing up shaped their brains physically in a new and different way (Prensky, 2001b).

Just to show to what extent marketing discourse and educational discourse are linked together, we could notice that "Digital Natives" and "Digital Immigrants" labels have been quickly adopted by marketing researchers to update and reinterpret the "old" generational targets, where the "Boomers" are to be considered as "Immigrants", the "X" as "Adaptive" and the "Y" and "Z" as the true "Digital Native" generations. On the other side, an intergovernmental organization like OECD (Organization for Economic Co-operation and Development) and its Centre for Educational Research and Innovation (CERI) promote the New Millennium Learners (NML) project focused on "the emergence of digital native learners" and aimed to "analyse this new generation of learners and understand their expectations and attitudes"(OECD). NML, from Howe and Strauss' label of *Millennials* (Strauss, & Howe, 1991, 2000), are outlined with the same words of Tapscott and Prensky, as "the first generation to grow up surrounded by digital media, and most of their activities dealing with peer-to-peer communication and knowledge management, in the broadest sense, are mediated by these technologies. Accordingly, Millennials are thought to be adept with computers, creative with technology and, above all, highly skilled at multitasking in a world where ubiquitous connections are taken for granted" (Pedrò, 2006).

2.1 Questioning the Net Generation

As I previously stated at the beginning of this article, I am not interested here in investigating the actual consistency of such a "Net Generation"; rather, I am interested I the theorical question about the role played by ICTs in its birth. How are they seem to affect the new generation? How do they enter in the generational identity-making processes? Are they the sole force shaping this identity? And how does this interpretation fit with the outcome of our "Media and Generations" project?

Some arguments come from the educational debate: David Buckingham, for instance, criticizing Tapscott's approach, stresses its technological determinism: "from this perspective, technology is seen to emerge from a neutral process of scientific research and development, rather than from the interplay of complex social, economic and political force. It is then seen to have effects...irrespective of the ways in which it is used, and of the social contexts and processes into which it enters" (Buckingham & Willet, 2006). Technology seems to be undetermined by any social force, but able to determine -by itself- the social change. In Buckingham's arguments there are some remarks useful to give further details about the conditions of generational building: on the one side, we cannot ignore the fundamental continuities and interdependencies between new media and old media "at the level of form and contents, as well as in terms of economics" (Buckingham & Willet, 2006); old and new technologies often turn out to co-exist, acting together upon (and being used together by) old and new generations, without splitting them into two "worlds apart". On the other side, "Tapscott's approach is also bound to ignore what one can only call the *banality* of much new media use" (Buckingham & Willet, 2006); it is not just a matter of the diffusion of a technology, that has to be so wide to be acknowledged as a common feature in the generational "we-sense": it is also a matter of the more common uses of that technology that can contribute to shape both the technology development and its impact upon its users.

More generally speaking, the "generational hypothesis" of the ICTs seem to be founded on an oversimplification of the processes shaping a generation, strongly focused on cognitive patterns (e.g. multitasking, networking, playing for learning and so on); very little is said about contents and imageries, that are often "old media" related even in the web (Hartmann, 2001); social and historical events seem to have no place; changes in the very ICT system, and in the broader media system, are not taken into account. Above all, these approaches seem to completely overlook the subjective dimension of generation-building and the endogenous forces operating *inside* the generation and *between* its members. In the very same way, "pure" generational marketing approaches seem to have a definition of generation that includes only three factors: it is a group of people who share the same life-cycle stage; live through the same economic, educational and technological times; were shaped by the same social markers and events. Such a concept of generation seems to be more the passive product of outer forces that are pushed upon the peer's group, shaping its values, its ideals (or lack of ideals), its *weltanschauung*, and determining, nowadays, its behavior, than a social and cultural identity. To sum up, both these approaches tend to press flat a very dynamic process in a sort of mechanical imprinting.

3 Conclusions: Memory, Space and Reflexivity

Prensky is probably affected by a form of "rhetoric of technological innovation" when he states that the arrivaland rapid dissemination of digital technology constitute a "singularity", a really big discontinuity "which changes things so fundamentally that there is absolutely no going back" (Prensky, 2001), but he's obviously right in pointing out the novelty of ICTs. If this discontinuity can hardly be seen as the only cause of the birth of a new generation, however it is possible to evaluate the impact of ICTs comparing them with the traditional media involved in the processes of generational identity-making I attempted to describe above.

My last point, in fact, is that the novelty of ICTs, with regard to the generations, lies in the transformations of these processes; these transformations, that affect in a peculiar way the roles played by the media, concern memory, space, and reflexivity.

Memory: according to the sociological tradition on generations as cultural identities, personal memory has a central role because it works like a stratification of experiences, facts and cultural meanings that characterise in the same way only the people who are the same age. In an "analogical" culture, the majority of the symbolic material used to build a common identity (the *repertoires* we registered during our research, made of remembered movies, TV programmes and so on) is ephemeral; there are, obviously, archives and libraries to conserve films, tapes and papers, but these institutions keep them behind closed doors and hide them out of the social visibility. Narratives, character and music survive in a very peculiar way in the stratified memories of the persons who encountered them at a certain moment of their life, when they were broadcasted or distributed; "having" them in one's own memory is a sure indicator of age and generational belonging and, for this reason, remembering them (or reminding each other about them) is a strong mechanism of the generational discourses.

On the contrary, evergreen and classics are always available to different generations, and for this reason they are often a link between the generations. ICTs and digitalization change the status of many cultural products, that can find a new repository in the Net, where everything seems to stand side by side with everything, in a timeless condition, always available to everyone. It is not just a matter of a new kind of preservative institutions: it is a new form of continuity of cultural provisions that works against disruption and discontinuity (and their power to mark the boundaries of elapsing time). With regard to the proposed model, it is likely "Media 1" are translated in "Media 3", losing uniqueness (and value) for their first audiences and becoming a resource addressed, for example, to a certain age group without any generational meaning (like classic cartoons of Disney, VHS pioneer of this transformation, or the re-runs of old TV programmes).

Space: in the sociological tradition, generations have a national dimension; traditional media and analogical technologies as well. National supply of cultural products, based on national languages, industries, deliveries and markets has been a constraint that conditioned the members of the past generations to use (more or less) the same, few, materials to build their own semantics. ICTs and digitalization act with no regards for national frontiers, redefining social space on both a global and a local dimension. What consequences are there for generational identities and belonging? Edmunds and Turner (Edmunds & Turner, 2005), for example, studying collective movements and political participation, suggest that, if that of the 1960s was the first global generation thanks to the new electronic media, nowadays Internet has created the basic technological conditions for the emergence of new global generations, linking together people through shared international experiences, introducing new reasons for political action and new means of coordination and activism, opening to new possibilities of sharing the same traumatic events, like 9/11 or Madrid and London bombings, with their capacity for providing "a focal point around which memories and political activism hinge" (Edmunds & Turner, 2005).

On the other hand, Volkmer, studying with a cross-national and cross-generational approach the shared consciousness of the global public space developed by different generations as a part of their own *weltanschauung*, discovered that depending on different media (radio, black and white TV and Internet) in the formative years different perceptions of the world tend to be created: while the oldest generations can be described as "place-based, where physical presence and identity are bound to a geographical life-world context"(Volkmer, 2006), the youngest seem to be "space-based", and "distance" and "proximity" are concepts defined in a very individual way. This way, ICTs increase the possibility for niche audiences, cultists and fans to share their symbolic material all over the world, using them to build a common cultural identity, at the same time emancipated from national binds and open to a global perspective.

But ICTs redefine social space giving more consistency to the local too; social networks maintain face to face relations, mobile phone is used for local coordination, some kinds of file sharing sustain personal friendship; concrete social groups can find an environment on the Net to enforce their shared identities, as well as the diasporic audience and users can build, thanks to virtual relations, a common identity, just like the "generation units" suggested by Mannheim.

Reflexivity: finally, this redefinition of the social space has consequences on the significant sphere of reflexivity and subjectivity. ICTs provide generations' members with a wide range of discursive resources where self-narrations can be told and self-representations can be acted. Blogs and UGCs are stages where generational semantics (and aesthetics) can be verified, proposed and contested. If the "big talk" of the Internet doesn't automatically mean an higher degree of self-consciousness or a more responsible way to participate in the "public sphere", it does show a wider availability of social visibility, and a tendency to use this visibility to express oneself in front of the "others".

This easy access to a "public sphere" is obviously very far from the opportunities given in an analogical culture to the members of the former generations, when conquering the social role that allowed to "speak on behalf of", or "in front of", the other members depended on matter of age and leadership. From this point of view, ICTs are likely to anticipate the processes of generational identity-making, taking them away from institutional constraints and professional hierarchy.

At the same time, these processes happen in a space densely populated by people belonging to other generations, intent on continuing on their own representations and to represent the "others". According to Hartmann, the "web generation discourse" is likely to be more a representation labelling the youth developed by the adults than a label the young use to represent themselves: if an Internet generation really exists, it is an "unwilling" generation, radically "digital" in ICTs practices but not in contents, not always conscious in terms of their actual use but aware of the general discourse that construct it this way. "Thus, there is a web generation, but only in the experience of the discourse rather than the actual experience" (Hartmann, 2001).

The transformations that affect memory, space and reflexivity because of the increasing importance of ICTs in the processes of generational identity-making supported by the media seem to me to have introduced some elements of problematicity; rather than becoming more evident and clear-cut, they become smoothed and softened; instead of cutting dramatically one generation from another, they are likely to link them; national boundaries fade away, and generations are able to build their semantics both on a local and global basis; the time needed for a generation to be born and develop its identity seems to shrink; but, at the same time, some aspects of generational identities largely diverge. Considering the generations in a multidimensional mode could be a way to once more derive some benefits from this concept.

Bibliography

Alexander, J.C., et al. 2006. *Cultural Trauma and Collective Identity*, Berkeley: Los Angeles: University of California Press.

Bourdieu, P., 1979, *La distinction. Critique sociale du jugement*, Paris: Minuit.

Bourdieu, P., 1988. *Homo Academicus*, Paris: Minuit.

Buckingham, D., Willet, R. 2006. *Digital Generations. Children, Young People and New Media*, London: Lawrence Erlbaum Associates

Corsten, M., 1999. The Time of Generations. *Time and Society*, 8 (2), pp. 249-272.

Dandaneau, S. 2001, *Taking It Big. Developing Sociological Consciousness in Postmodern Times*. Thousand Oaks: Sage.

Edmunds, J., Turner B., 2002. *Generations, Culture and Society*, Buckingham Open University Press.

Edmunds, J., Turner B., 2005. Global Generations: Social Change in The Twentieth Century. *The British Journal of Sociology*, 56(4), pp. 559-577.

Elder, G.H., 1974. *Children of the Great Depression. Social Change in Life Experience*, Chicago: University of Chicago Press.

Eyerman, R., Turner, B., 1998. Outline of a Theory of Generations. *European Journal of Social Theory*, 1(1), pp. 91-106.

Hartmann, M., 2003.*The Web Generation? The (De)Construction of Users, Morals and Consumption*, Brussels: SMIT-VUB, Free University of Brussels.

Heide, M., 1995. *Television Culture and Women's Lives. Thirty something and the Contraddiction of Gender*, Philadelphia: University of Pennsylvania Press.

Kortti, J., Mähönen, A., 2009. Media Ethnography, Oral History and Television. *European Journal of Communication*, 24(1), pp. 49-67.

Koselleck, R., 1969. *Kritik und Krise*, Freiburg.

Mannheim, K., 1952. The Problem of Generations. In: Mannheim, Karl (edited by Paul Kecskemeti): *Essays on the Sociology of Knowledge*. London: Routledge and Kegan Paul pp. 276-320.

Meyrowitz, J., 1985. *No Sense of Place. The Impact of Electronic Media on Social Behavior*, New York, Oxford: Oxford University Press.

Moores, S., 1988. The Box on the Dresser: Memories of Early Radio. *Media, Culture and Society*, 10(1), pp. 23-40.

OECD. Avalaible at: http://www.Oecd.Org/Document/10/0,3343,En_2649_35 845581_38358154_1_1_1_1,00.Html [Accessed 8 May 2009]

O'Sullivan, T., 1991. Television Memories and Cultures Of Viewing, 1950-65. In: J. Corner, ed. *Popular Television in Britain. Studies In Cultural History*, London: Routledge.

Pedrò, F., 2006. The New Millennium Learners. Challenging our Views on ICT and Learning. Available at: http://www.Oecd.Org/Dataoecd/1/1/38358359.Pdf. [Accessed 8 May 2009].

Pilcher, J., 1994. Mannheim's Sociology of Generations: an Undervalued Legacy. *British Journal of Sociology*, 45, pp. 481-495.

Pinder, W., 1926. *Das Problem Der Generation in Der Kunstgeschichte*, Berlin; Frankfurter Verlaganstalt.

Prensky, M., 2006. Digital Natives, Digital Immigrants Part1. *On The Horizon*, 9(5), pp.1-6.

Prensky, M., 2001. Digital Natives, Digital Immigrants Part2. Do They Really Think Differently ? *On the Horizon*, 9(6), pp. 1-6.

Press, A., 1991.*Women Watching Television. Gender, Class, and Generation in the American Television Experience*, Philadelphia: University of Pennsylvania Press.

Silverstone, R., 1994. *Television and Everyday Life*, London: Routledge.

Silverstone, R., Hirsch, E., Morley, D., 1992. *Consuming Technologies: Media and Information in Domestic Space*, London: Routledge.

Smith, J., Clurman, 1997. *A. Rocking the Ages. The Yankelovich Report on Generational Marketing*, New York: Harper.

Strauss, W., Howe, N., 1991. *Generations: The History of America's Future, 1584 To 2069*, New York: Vintage.

Strauss, W., Howe, N., 2000. *Millennials Rising: the Next Great Generation*, New York: Vintage.

Tapscott, D., 1998. *Growing up Digital*. New York: McGraw-Hill.

Tapscott, D., 2008. *Grown up Digital. How the Net Generation Is Changing your World*, New York: McGraw-Hill.

Twenge, J.M., 2006. *Generation Me*, New York.

Volkmer, I. ed., 2006. *News In Public Memory. An International Study of Media Memories Across Generations*, New York.

Wyatt, D., 1993. *Out of the Sixties. Storytelling And the Vietnam Generation*, Cambridge.

Jukka Kortti

The Problem of Generations and Media History

1 Introduction

The idea of generation as a sociological concept is problematic in many ways. As often with social theories, it simplifies and stereotypes multidimensional phenomena. Even though the sociological idea of generation is problematic in relation to the media – as well to the history of media – it can be a valuable approach in media research.[1]

The media often take an important role in the generational movement. They are both a device for creating generational identity and an arena for remoulding the image of generations. Certainly generation also has significance on the consumption of the media, particularly within the digital culture: different age groups use the media in different ways. However, in this article, I will focus specifically on how the media have functioned as the *voice* of generations and how they create the image of generations.

When talking about generations and media history, I am particularly interested in what kind of role the media has in creating generational consciousness – actively and reflectively or on a more subconscious level. I will also depict how the media are shaping the image of a generation – how the discourse of a generation is realised through the media. I would like to question or at least remould and define the sociological ideas of generations in historical studies. I emphasise the concept of the idea-historical world view or philosophy of life, in particular, through which it is easier to access the thematic structure of generation.

My examples mainly come from my current research into the 100-year history of the student magazine *Ylioppilaslehti* (University Student Magazine). *Ylioppilaslehti*, which is still published today, was founded in 1913 as the magazine for all Finnish-speaking regional student associations – so-called student nations. However, *Ylioppilaslehti* is not just "any student paper"; it is a significant Finnish cultural and political institution. Through its editors, it bore has borne witness to the lives of the majority of the Finnish political and cultural intelligentsia of the 20th century. The reason for using the theorising of genera-

---

[1] For general reflections on the concept of generation and the different disciplines that approached it, see Colombo, infra.

tions in my research is its, some might say exceptional, applicability for media-historical analysis.

My samples concentrate on the 1920s and 1930s on the one hand and the 1960s and 1970s (particularly 1968) on the other. Both periods were particularly interesting in the context of modernism which is crucial for the theories of sociological generations. As in many European countries, Finland also had (Fascist) right-wing radicalism in the inter-war period and new left radicalism in the 1960s and the 1970s. Media played an important role in both periods of generational movements.

In addition to printed media, my examples also come from Finnish television. Its most intense period of growth and unequalled significance as a means for human interaction took place in the 1960s – even though in the United States the era of television which revolutionised society and culture had already begun a decade earlier. Television was the Internet of the 1960s: "revolutionary mass media" and according to Macluhanism "the extension of man". Additionally, I refer to the events and the media of other, larger countries during the same periods. In the 1960s, in particular, the generational movement was already very international but even at the beginning of the 20th century the generational movements of different countries had many characteristics in common which were differed according to the specific national situations.

2 Modern, Elite Generations

Generation as a concept is a fairly new one. The roots and prehistory of societal generation concept can be found in the 19th century. Views and theories concerning the generations of this period were united by the withdrawal from the ancient genealogical meaning: generations were now considered on the collective level and not only as being completely analogous to family generations. The significance of generations as a social phenomenon was emphasised in general. However, views from the 19th century did not yet include the idea of experience (of youth) uniting generation members and the generational consciousness produced by it. Generations were still mainly regarded as age groups, even though as the century progressed it began to take on its modern meaning. Auguste Comte, John Stuart Mill and Wilhelm Dilthey were three of the most important writers to publish work about the generations of the 19th century and they created new viewpoints compared with the previous thoughts for understanding generations. (Purhonen, 2007, p.23).

All of the most important modern generation theories were formulated in Europe between 1910 and 1933, particularly in the 1920s. The Germans mainly produced ambitious societal theories on the concept of generations, whereas the English wrote poems and novels which articulated the conscience and destiny of generations. French descriptions of the characteristics of generations were created by different organised writer groups and the Italians, first and foremost, analysed generations in their political essays (Wohl, 1979, pp.2-3.)

This group of generation theoreticians from the beginning of the 20th century includes the Spanish cultural philosopher José Ortega y Gasset, the German art historian Wilhelm Pinder, the French cultural philosopher François Mentré and the German sociologist Karl Mannheim. The common factor in their theories was that their conclusions were more or less based on the ideology of the modern, educated middle class elite living its youth. The main problem when applying generation theories is indeed that often any age cohort in its entirety is interpreted using a theory-based approach. In other words, the activities of the small, loud vanguard of youth are identified with the entire coeval masses.

Despite this, or in fact as a result of this, classical generation theories can often be used in media research. The media have often had an important role in generational movements. This is emphasised in the history of media such as *Ylioppilaslehti*. Firstly, the writers of *Ylioppilaslehti* have always been a small elite. Secondly, academic youth fit naturally into the generation concept because we explicitly talk about youth being at a certain 'sensitive age'.

However, generation theories are not appropriate for all times and places. When used in historical research, the concept of generation is related to a problem which could be called the vantage point of modernism. The societal generation is very much a concept of the modern thinking based on the Enlightenment. It includes defining the status of youth (generational categorisation) and the membership of a national state: nationality. Nationalism connected with generational movements – particularly at the beginning of the 20th century – has often been considered to be created, first and foremost, by modernism (Anderson, 1983). Additionally, many "children of the Enlightenment", such as liberalism and socialism have emphasised a difference to the past.

Therefore, the generation theorising first attracted widespread attention in the 1920s which can be regarded as the beginning of a period of culmination of 20th-century modernism as regards economy, culture, and art in particular (Futurism, for example). The next time the concept of generation was particularly present was in the 1960s which can be regarded in many ways as the end of modernism or at least as its period of redefinition.

A number of significant historical situations triggering radical changes preceded both periods. Hence the formation of a generation depends upon the relative speed of social change because when social change is very slow it is difficult to distinguish between generations. Alternatively, when social change is very rapid, generational differences are magnified. Emergence of a new generation depends entirely on the trigger action of the social and cultural processes (Rintala, 1979, pp. 8-9; Mannheim, 1972)[2].

These societal changes which have an effect on youth, in particular, are often called *key* or *formative* experiences by generation theoreticians. As concepts they are similar to those of Karl Mannheim (1972, p.120) who writes about formative tendencies and formative principles which can bind groups together and which are capable of becoming the basis of continuing practice. These formative experiences and events may include wars, such as World War I in Robert Wohl's (1979) generational study. WWI remoulded the European inter-war intelligentsia, some of whom became communists or liberals and some of whom became fascists. The formative exepierence can also be a series of significant events, such as the Vietnam War, demonstrations and the occupation of universities in 1968. Often these are connected with strong modernisation processes within society. As Mark Kurlansky wrote:

> "1968 was a time of shocking modernism...and modernism always fascinates the young and perplexes the old., yet in retrospect it was a time of an almost quaint innocence."(Kurlansky, (2004, p. xix)

3 Zeitgeist and Traditions

The most famous of the generation theories of the 1920s is undoubtedly Karl Mannheim's article *Das Problem der Generationen* from 1927. The theory has proved to be particularly useful when researching the previously mentioned periods in which youth played an important role. Kurlansky (2004, p.xix) added to the previous quote, "modernism always fascinates the young and perplexes the old". Young people (Jugend, Giovinezza), in general, were at the forefront of movements and generation theories at the beginning of the 20th century.

Mannheim also has something to offer to studies focusing on the media. For media-historical research, it is essential to define how generations are examined, first and foremost, as historical phenomena of their own era. This also clarifies the role of the media within generational movements. Here, I approach Mannheim's redefinition from the starting point of one of his essential theoretical apparatus.

Zeitgeist (the spirit of the age), already used by German Romanticists and Hegel, was an important element of Mannheim's thinking in his generational

---

2 In some countries modernisation processes have been more drastic than in others. A common factor for Finland and Italy, for example, has been the speed of the modernisation process at the European scale both at the beginning of the 20th century as well as in the 1960s.

theorising,[3] Zeitgeist was also introduced to historical research in the 19th century by history theoretician Leopold von Ranke and others from the German school of empiricists. It sprang from the need to separate the natures of the past and present. The spirit of the age referred to the presumptions and unquestioned basic experiences of a certain historical era. In extreme historicism this may have meant that some epochs were incommensurable because their essences were so different (Davies, 2003, pp. 29-30).

However, continuum had significant meaning in Mannheim's concept Zeitgeist – how strongly tradition impacts on the actualisation and mobilisation of generations. In a way, generations create a new version of tradition or at least it is very much involved in forming the spirit of the new age. Mannheim's example was the German generations of the 1830s which had their roots in romantic-conservative tradition. It was remoulded by one generation unit but at the same time, a liberal-rationalistic group appeared which moulded tradition in a completely different way, even the opposite compared with the other unit. (Mannheim, 1972, pp.130-131). In this way, different generation units were born.

Remoulding of the previously mentioned tradition for the Finnish academic youth took place in the 1920s and the 1930s, in particular. During its first years, *Ylioppilaslehti* was mainly a "professional magazine for students" and it reflected the general non-political sentiments of the student world. However, the situation changed in the 1920s. The activities of the university students concentrated specifically on the Academic Karelian Society (AKS). The first institution which the AKS took over within student circles in 1923 was *Ylioppilaslehti*. The AKS was an extreme right-wing student movement founded in the previous year[4]. All in all, *Ylioppilaslehti* was a strictly nationalistic Finnish-minded propaganda magazine throughout the 1920s and 1930s. In this sense, as a thoroughly political university student magazine, it was exceptional even when compared to the international arena.

In the AKS, there was also a strong belief in the national spirit, the historical mission of a nation. This way of thinking originated from the idealism of Romanticism and Hegel's national philosophy. The thought behind Hegel's *Blut und Boden* (Blood and Heritage) which could be found in the Fascist movements of Europe of the era, was also a part of the state socialistic ideology of Yrjö Ruuth, the ideological teacher or master of the AKS. The aim of AKS was to strengthen Greater Finland, both externally and internally.

---

3 Zeitgeist was extensively a part of Mannheim's sociology of knowledge.
4 The association emerged from the revenge-spirited Karelian idea. The original main idea was to get back the Eastern Karelian parts left in Soviet Russia in the "Shame Treaty" of Tartu. According to the treaty Finnish troops were to be withdrawn from two large border parishes of East Karelia, Repola and Porajärvi, which had been occupied by Finnish troops since 1918 – after Finland became independent from Russia in 1917.

The AKS was "the heir of the Fennoman movement" as Matti Virtanen, who has researched Finnish political generations according to Mannheimism, interpreted it. The Fennomans were the most important political movement in the Grand Duchy of Finland in the 19th century. The movement pushed to raise the Finnish language and Finnic culture from its peasant-status to the position of a national language and national culture. Virtanen also considers the generation (or generations) of the 60's and *Taistoism* born from it as the heritage of the Fennoman movement. Taistoism was a Finnish version of the Stalinistic, orthodox pro-Soviet Eurocommunist movement in the 1970s. According to Virtanen, the new left movement went through different phases but actually originated from the Fennoman movement. It was only actualised differently. (Virtanen, 2001, pp.380-381).

In my opinion, explaining academic generations as hegemonies of different spirits of age based on a certain tradition, which are actualised in certain ways and times is too vague. When properly adapted, it explains everything and it actually makes the concept of generation unnecessary in this sense. In order to better examine this essential concept of generation and the role of the media in it, we need to focus on the theory of Zeitgeist.

4 Zeitgeist or Philosophy of Life?

In Mannheim's view, the same era is different for people of varying ages and thus experiencing the spirit of a common era is only achieved between coevals. Therefore, eras cannot be united by one and the same spirit of time, consequently a generation is only left with the spirit. Even though Mannheim resisted the view of one total spirit of time, he still considered that the differing views of different groups would eventually form one united spirit of time. Thus he thought, unlike Pinder, that different and even opposite generation units will eventually articulate destiny, uniting the whole generation. The spirit of time for Mannheim was the "spirit of an epoch" or a "mentality of a period". (Mannheim, 1972, p.129; Purhonen, 2007, p.42).

Even though Zeitgeist is not a synonym of mentality, generational research could be taken further via the concept of *mentality* guided by Mannheim. In this way, the position of media in the generational movement can also be located more clearly. Robert Wohl (1979, p.4), for example, who researched European "generations of 1914", discusses mentalities, both individual and collective. As he writes, much of his writing concerns "the intuitions, feelings, and ideas of males from the middle layers of society".

Mentality includes different thought patterns, norms, mores and traditions. Mentality absorbs substance from both thoughts and actions. In fact, it focuses specifically on the relationship between thinking and action. The subconscious behaviour of everyday "normal life" in particular, is the main object of interest. Mentality is a means of structuring and perceiveing the world – a mental map where one's understanding of reality can be located. In historical research the history of mentalities focuses on conceptual prerequisites for attitude or stance and how they have changed.

The advantage of an approach to the history of mentalities relation to the history of ideas, for example, is that it is also concerned with the often subconscious motives of individuals and groups. As cultural historian Peter Burke (1986, p.439) says, the history of mentalities was born to fill the gap between the history of ideas and social history, being a certain historical anthropology of ideas.

However, the history of mentalities as a viewpoint is too broad for the interpretation of generations. As stated previously, societal generation – at least in the Mannheimian sense and therefore also in this context – explains elite groups in particular. Therefore, thinking that the generation's spirit of time consists of a vast ensemble, which includes the ensemble of feelings and mores, takes the interpretation of generations beyond of its core (or should I say under its core). As a concept, mentality is therefore much more extensive than Zeitgeis*t* but also within this concept one can find the same problems that are found in Zeitgeist: it encourages historians to over-estimate the intelligent consensus of past societies (Burke, 1986, p.443).

One can generalise by saying that a mentality-historical approach is suitable for so-called 'ordinary' people who do not include the academic elite or journalists, for example. The history of mentalities is more than ideas and ideologies. Even though the mobilisation and activities of generations are related to different mores, and undoubtedly also to feelings, it would be preferable to try and leave them out conceptually when analysing generations. The history of mentalities often locks conscious thinking into the room of the history of ideas. Mentality can be seen as part of the *world view*. Finnish historian of ideas Juha Manninen (1977) defines the world view as a practice of life: the world view is manifested in the peoples's relationship to their past and future, their possibilities, their personal and societal life etc.

This depicts the actualised academic generations rather well. The past-future dichotomy was essential to both right-wing radicals of the AKS and left-wing Taistoist radicals– regarding how the previous generations were regarded and how there would be a perfecting future ahead. It also becomes evident how the world view is connected with information retrieval, which is often provided by the media.

The world view also includes conscious "striving to systematise the views of reality", not only "by silencing acquired societal, cultural and conceptual prerequisites" as Manninen states. The world view is not separate from societal interests but also requires people and individuality. Therefore the world view does not either revert directly to material group or class benefits. (Manninen, 1989a, p. 41; Manninen, 1989b, pp. 227-228.)

Even though academic movements have unquestionably strived for systematisation, the radical right-wing AKS, for example, was labelled as a strong group and class point of view (peasantry), particularly in the 1930s, not to mention the radical left-wing Taistoists for whom the working class was an essential framework[5]. The concept of world view as such would be more suitable than the concept of mentality to be used in connection with societal generations, but it seems too vague. In fact, the world view includes mentality. However, the concept of generation can be examined from the perspective of world view, if we make it more specific by dividing it into two.

World view, which includes mentality, does not in its entirety require mental activity from a subject. The concept of *philosophy of life* (Weltanschauung) depicts this active side. The philosophy of life is a certain reflected dimension of the world view. It requires exceptional mental activity from an individual; it involves striving to systematise views on reality. This is close to the next sibling concept; ideology. However, ideology is different from philosophy of life as it is always intended for several individuals, whereas philosophy of life can also be understood on an individual level. (Manninen, 1977, pp.22-26; Manninen, 1989b, pp.227-228; Manninen, 1987, pp.136-137).

Thus world view could be explained in such a way that mentality is the subconscious dimension and philosophy of life is the conscious dimension of the world view. If the societal idea of generation focused on the aspect of the philosophy of life, the analysis of generations could depict the essence of generations more clearly. However, the precondition for this approach would be to research "elite generations", such as academic youth where individuals – in practice the leaders of generational movements – hold an important position concerning the actions of the movement.This can be illustrated as follows:

---

5   On the other hand, generational theories have also been considered as an alternative for class theories (Wohl, 1979, pp. 81–82).

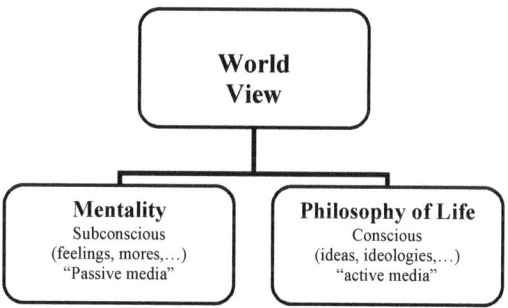

Figure: Media and World View (Source: the author)

However, this does not mean that research would be an idea-historical interpretation which would only focus on thoughts conceived by great men. Other more extensive entities, including subconscious mentality-historical ones, are introduced by narrating the required context. The history of mentality always requires a comprehensive viewpoint both when reading source material and when understanding the culture to be researched. The mentality-historical approach in depicting a context is in any case important in order to avoid anachronisms which often occur with the history of ideas: the intellectual world of generations must be seen as part of its own era and from that point of view.

A good example of this is to introduce the relevant popular culture of the time, or particularly the media, to give background to a particular some ideologically-loaded movement, through which the world view of a generation units is created. Thus, the media operates on both subconscious and conscious levels: it is part of the mental-historical context often linked with culture which has an effect on an individual or it may be a device for creating a philosophy of life. The "active media" governed by a generation unit also remoulds the "passive media" of the mentality level but the opposite effect is also true. The media can also abuse the image which the generation has created of itself.

5 The World is Viewed Through Bifocal Glasses: Two Student Leaders from the 1920s and the 1930s

I have applied the previously mentioned theory apparatus to two Finnish student activists and chief editors of *Ylioppilaslehti*; Vilho Helanen and Urho Kekkonen. With this theory I particularly want to prove how the media is utilised during a certain phase in the development of a philosophy of life of generational movement leaders, which either works as the basis for moulding a philosophy of life later in their lives or for not changing past habits.

Both of these charismatic student leaders were excellent rhetoricians and skilled writers. At the beginning of AKS in the 1920s they were the organisation's "journeymen"[6]. However, after the movement split in 1932, Helanen became the undisputed master of the AKS. The group of first-generation AKS members resigned as they did not accept policies which were becoming Fascist. Although only just over one tenth of the members resigned, they included the most renowned AKS members and several former chief editors of *Ylioppilaslehti*.

Later Kekkonen, the 8th president of Finland between the years of 1956-1982, became the Great Master of the 60s generation. The formative experience of Kekkonen's generation (born 1900) was the Finnish Civil War in 1918. Therefore Kekkonen was at exactly the right age for a formative experience. According to Mannheim (1972, p.115), the age of seventeen is important for later development and he states: "The possibility of really questioning and reflecting on things only emerges at the point where personal experimentation with life begins – round about the age of 17, sometimes a little earlier and sometimes a little later."

Helanen's formative experience (born 1899) was not so much the Civil War, which he missed, as the Estonian War of Independence[7] which he took part in as a volunteer. Kindred spirit stemmed from there and Russophobia caught fire. Russophobia concerned the whole AKS generation born at the turn of the century, including Kekkonen, whose school years were during the darkest Russification period; experiences from Russia were nothing but bad. Helen Sepp (1997,

---

[6] The masters, journeymen and apprentice model, launched by Finnish social scientist Matti Virtanen (2001, pp.351-390), is fairly usable for the academic youth. Virtanen's idea, in brief, is that the oldest and the most experienced faction of the generational tradition forms the class of masters, the journeymen form the middle layer and the latest arrivals become apprentices. Often academic youth has been led by slightly older activists who have received their education from an older ideology or a thinker. Those in the age of the formative experience are particularly prone to the media influences delivered by the older elite. The same idea that the older generation always educates its successor can also be found with Antonio Gramschi.

[7] The Estonian War of Independence (Estonian: *Vabadussõda*, literally 'Freedom War'), was a defensive campaign by the Estonian Army and its allied White Russian Northwestern Army against the Soviet Western Front offensive and the Baltic German Landeswhr offensives in 1918-1920 in connection with the Russian Civil War. The campaign was the struggle of the Republic of Estonia for sovereignty in the aftermath of World War I. It resulted in a victory for Estonia and concluded in the Tartu Peace Treaty. Through the Russian revolution, both Finnish Civil War and the Estonian War of Independence were the consequences of World War I.

p.49), who has researched Helanen, calls this age class the generation of the Russification of Finland[8].

Helanen's philosophy of life resembled one from the 19th century, a national romantic and comprehensively synthetic view familiar to the Fennoman movement. As an idealist trusting intuition he blindly believed in his own vision which he practised passionately and unpredictably. Even though this depicts Vilho Helanen, above all, it can be connected for the most part to the mentality of the whole AKS elite of the same generation in the 1920s. How did the philosophies of life then part when at the start of the 1930s?

Still in the early days of the Fascist Lapua Movement[9] at the turn of the 1930s, the whole of AKS joined the activities of the movement. Communists were seen as Soviet Russia's Trojan horse and destructive for the fatherland. The most important difference was that Helanen's boot of idealism was ready to trample upon legality if necessary, whereas the more moderate members of the Centre, such as the lawyer Urho Kekkonen, definitely wanted to remain within parliamentarism and legality. At the same time Helanen's ideal of integration was bending and the workers were no longer part of the future Greater Finland.

Conflicts concerning AKS's policy then came to a head with two former chief editors of *Ylioppilaslehti*, Helanen and Kekkonen. As students in 1924, they had participated in the most visible AKS performances together. Still in their chief editor periods, their versions of *Ylioppilaslehti* (Helanen as chief editor in 1925, Kekkonen in 1927) were quite similar to one another. Both magazines addressed many issues concerning the AKS, Finnishness and kindred spirit. Russians and Swedes were regarded with hostility. Kekkonen and Helanen both got their claws into the competing radical right faction, Independence League, but they wanted to integrate social democrats into Finnishness.

---

8   The Russification of Finland (1899-1905, 1908-1917, *sortokaudet/sortovuodet* (times/years of oppression) in Finnish) was a governmental policy by the Russian Empire aimed at terminating the autonomy of Finland. It was part of a wider policy of Russification pursued by late 19th-early 20th century Russian governments which tried to abolish cultural and administrative autonomy of non-Russian minorities within the empire. In his classic study on nationalism, *Imagined Communities*, historian Benedict Anderson (1983) considers that "Russification" refers to all official nationalisms of superpowers where nationalities of an imperium are intended to be assimilated into the system of the mother country, for example.

9   The Lapua Movement (Finnish: *Lapuan liike*), sometimes referred to as "Lapua Fascism", started in 1929 and was initially dominated by ardent anti-communist nationalists, emphasising the legacy of nationalist activism, the White Guards and the Civil War in Finland. The movement saw itself as the badly needed restorator of what was won in the Civil War. It radicalised and turned into more of a Fascist movement but was banned after a failed coup-d'état in 1932. The activities were then continued with the Fascist People's Patriotic Movement (IKL).

The only small differences in the content of the magazine were that during Kekkonen's period there were slightly more culture and sports-oriented topics. Helanen's magazine, for its part, had more temperance and church-related topics. Perhaps by considering their philosphies of life, one can simplify slightly: Helanen could be seen as a religious man wanted to enlighten, while Kekkonen could be seen as a sportsman[10] who wanted to compete.

It was essential for their philosophies of life that in the "AKS school" of the 1920s Kekkonen was strongly influenced by the AKS master and theoretician Yrjö Ruutu, whereas Helanen was more influenced by another master, Elmo Kaila, who was more oriented towards practice. This development becomes evident when considering their numerous writings in different forums. As emotional people, theories were secondary for Helanen whereas Kekkonen was fascinated by Ruuth's ideas of State socialism concerning the independent role of the middle class between capitalists and workers.

Kekkonen became Helanen's most important sparring partner in the 1930s. During the split of AKS, Kekkonen was writing his dissertation in Germany. However, he was one of the main architects of the separation. Close relations to National Socialism made Kekkonen feel antipathy towards the radical right and Finnish fascism, too. On the other hand, Helanen also despised the later Estonian politics of the Germans and the fact that Nazism opposed Christianity. Kekkonen also began to draft his conception of democracy which became more concrete in his numerous newspaper writings and the book entitled *Demokratian itsepuolustus* (Self-defence of Democracy) (1934). Relations between Kekkonen and Helanen were irreparably damaged when Kekkonen as Minister of the Interior wanted to outlaw the activities of the IKL (the People's Patriotic Movement founded from the ruins of the Lapua Movement) and to discontinue its journals. At that time, the AKS (led by Helanen) was already, in a sense, part of the Fascist IKL party.

The philosophies of life of two student activists and former chief editors of *Ylioppilaslehti* of the same generation differed in how their perspectives were directed after their time as chief editors of *Ylioppilaslehti*. Helanen looked at the world, first and foremost, through expired-ideal glasses which were intended for close reading. Helanen got "stuck" in directing all his energy towards educating the academic youth which he considered more essential than national politics, whereas Kekkonen already focused on a broader horizon with modern bifocal glasses. Helanen had to stay in the wilderness after the Second World War when the Fascist policy he represented was defeated. He was even condemned as a war criminal and in his final years, before he fell terminally ill, was mostly

---

10 Kekkonen was an active athlete. His best achievement as a sportsman was to become Finnish high jump champion (1.85 m) in 1924.

recognised as a writer of detective stories. Kekkonen, for his part, became the greatest Finnish statesman of all times.

Comparing these two student leaders shows how actors within similar experiential history, from the same generational movement and from the same media are later oriented towards different policies as a result of their philosophies of life. The societal and cultural mentality experienced by both was the same but the fact that their philosophies of life oriented towards different directions moulded their world view differently.

Often the directions of elitist generational movements are dependent on the actions of its leading figures rather than on units and factions which would explain everything. Generational movements need charismatic leaders and the media play an important role in this. Often the media constitute a device and a channel for cultivating view points concerning philosophies of life and generations. Both student leaders Kekkonen and Helanen were diligent writers and *Ylioppilaslehti* was only one of their media. However, Kekkonen's policy developed, whereas Helanen's philosophy of life played the same record – or should I say the same typewriter. Kekkonen changed as his philosophy of life developed he remoulded himself into a broader mentality.

The media can therefore be a voice of the philosophies of life for a generation which is used to create the basis for the later formation of a world view or it can be a device to which the same ideas are chained. When Zeitgeist changes in the latter case, the media do not correspond to the spirit of the age either. This also happened to *Ylioppilaslehti* when the post AKS-generation was forced to emphasise culture instead of politics immediately after the Second World War.

Situations such as this, where the generation leaders and ideologists separated from the same movement use the media to develop and spread their philosophy of life as a political and cultural project, can be found in several generational movements in Europe from the beginning of the 20th century. For example, *Leonardo* by Giuseppe Prezzolini and Giovanni Papini and particularly Prezzolini's later *La Voce* played an important role within generational movements in Italy at the beginning of the 20th century which emphasised youth and militancy. The same interventionist movement also produced the Liberal historian Adolfo Omodeo and the Futurist Filippo Tommaso Marinetti. Before World War I, leftism united people who would later be oriented towards very different directions, such as Benito Mussolini and Antonio Gramsci. For all of them World War I was in many ways a formative experience for their later thinking and journals were an important forum for developing ideas (Wohl, 1979, pp.160-202).

6 Battle for Hegemony

Gramsci also covered the concept of generation in his *Prison Notebooks* (*Quadermi del carcere*) and he perceived generations through the elite, in particular. However, he saw the members of the elite in generational movements mostly as "clerks" or "errand boys". They were those who realised *cultural hegemony* which is renownedly Gramsci's essential concept. One group of intellectuals which sustained the societal hegemony and political power of ruling social groups were newspaper editors. (Gramsci, 1996, pp.52-53; Wohl, 1979, pp.195-196).

It was also a matter of generational media hegemony in *Ylioppilaslehti* of the 1920s and the 1930s. The separated AKS faction tried to reach hegemony immediately after its separation in the Student Union right by promoting its own man for the position of Chairperson of the Student Union in 1932. The candidate of the separated faction lost by a whisker to the majority's candidate and this was influenced by the fact that *Ylioppilaslehti* was still ruled by the separated AKS faction.

The separated members founded a paper called *Ylioppilas* (University Student) which was published from 1932 to 1933. In 1932 the even more liberal *Vapaa ylioppilas* (Free University Student) was also published. All in all, the intense idealism of the 1930s created numerous student papers which wanted to provide an alternative for *Ylioppilaslehti* both from the left and the right and from other political stances. The editorial of *Ylioppilaslehti* (vol. 18B/1930, 375) commented on its competitors in 1930: "Far from belittling steady intentions of the previously mentioned papers to fill their position as mouthpieces of their interest clubs, we want to make clear that they are all born as an accomplishment of an active individual rather than being created from an actual need." Thus the importance of an individual's philosophy of life as a maker of factions was already acknowledged at that time.

The Finnish academic youth was not in any case united even in the 1930s. Even though the hegemony of the AKS remained within the environment of university students, it also had its active opposition. As Gramsci states, ideology must be seen as a battle field where two hegemonial principles are against each other. It must also be emphasised here that neither did Gramsci perceive ideology as individually gained (as the philosophy of life can be) but it is always adopted through the intermediary of an ideological terrain. (Mouffe, 1985, p.226) *Ylioppilaslehti* was one of the most important "intermediaries" of this kind for the AKS.

In British cultural studies, arising from the new leftism, the ideology theories of Antonio Gramsci and Louis Althusser had an important role and they were popular in communications research particularly in the 1970s and the 1980s. The interpretation of Gramsci's concept of hegemony by the "father" of British cultural research, Raymond Williams, was that hegemony is not static or systematic but a continuous historical process which changes constantly. Hegemony consisted of a cultural process which was formed of traditions, institutions and different formations. The mass media as an institution had an important role in this. (Stevenson, 1995, p.17.) It was, according to Althusser (2002), one of the "ideological state apparatuses".

Particularly in the generationally bound movements of the 1960s, it was a matter of *media hegemony* in many ways – of who ruled the media and how. It was not only about a political battle for supporters and power but of "who defined which cultural and societal issues were discussed and on whose conditions", as Tapani Suominen (1997, p.32), who has researched European new leftism, writes.

7 Television and 1968

It can be said that 1968 – "the year of the barricades" and "the year that rocked the world" (Caute, 1988; Kurlansky, 2004) – was also a milestone from this point of view. It was particularly a matter of the collision of traditions and institutions. Hegemony was questioned and the aim was to create a new – according to Gramsci – "collective national-popular will" (Gramsci, 1996, p. 62-63; Mouffe, 1985, p.232). For generation theories, the media had an even more significant and extensive role in the generation movement in the 1960s than in the period between the World Wars. In the 1960s the press was at the focal point of phenomena but the philosophy of life for youth was also constructed by new electronic media – television, in particular. Television was important especially on the previously described mental level but it was also active in creating philosophies of life.

Television had an important role in the movements of 1968 in three ways. Firstly, television rapidly broadened baby boomers' world view by reporting about worldwide evils by fast communication enabled by satellites and (professional) videotape. Above all, the reality of the third world, such as Biafra's famine catastrophe and Vietnam's "first television war", quickly awakened the feeling of collective injustice. Secondly, television very soon gave models on how to act and it created a coherent story on phenomena taking place around the world[11]. Thirdly, television provided an efficient channel for public sphere. Its

---

11 For example after the American radicals followed the occupation of Sorbonne University in television in May 1968, their own second occupation of Columbia University was different with its barricades from the first one at the beginning of the year. As Daniel Cohn-Bendit, the student leader of riots in Paris stated, they met other radicals of the world through television. This did not happen by direct personal relationships but with the images created by television. (Kurlansky, 2004, p.224.)

role was not only mentality-historical but it was also actively connected with philosophies of life. Fairly soon the movements of the time noticed that the more extravagantly they performed and clashed with the authorities, the more "the whole world was watching"[12]. This way television was also strongly creating the image of the generation of the 60s.

1968 was also significant in Finnish broadcasting policy; in defining policies for its future. The national broadcasting company Yleisradio was by that time discontinuing the commercial, commercially-funded Mainos-TV which was founded ten years earlier to fund Yleisradio and which was owned by business life. Mainos-TV was saved just in time[13]. Behind the discontinuation project was a new and modern idea of broadcasting policy which collided with that of the conservatives. In the late 1960s, *Reporadio* was a nickname given to the Finnish national broadcasting company after Director-General of Yleisradio Eino S. Repo (1965–1970). *Reporadio* is remembered, among other things, for its "informational programme policy" and radical, leftish ideas about public broadcasting.

To put it simply: the question was about who defines Finnish communications: the government or the business life, the right or the left, radical progressive baby boomers or middle-aged conservatives. In connection with commercial television, *bourgeoise hegemony* was often mentioned – it was a fashionable

---

12 Reference to Todd Gitlin's book The Whole World Is Watching: Mass Media in the Making And Unmaking of the Left (1980). According to the title, the book examines the media's role in the movements of the American New Left in the 1960s. Gitlin, who himself was one of the leaders of the American New Left, also wrote a more extensive book on the 60s a few years later (Gitlin, 1987). In this book, the media was involved more analytically than which was usual in research and in general presentations concerning the movements of the 60s.

13 Although Finland was quite late to start television broadcasting in the European context, there were some unique features in the Finnish development of television. First of all, Finland was a pioneer in the European commercial television. In Finland, the first television broadcasting company was a commercial enterprise which was started in early 1956 by technology students who had established their own station in 1955. Meanwhile, adversaries in the national debate over control of Finnish television reached a compromise. Like many other European nations, Finland saw strong advocacy for making television a public, non-commercial medium. In the end, however, the compromise arrangement patterned itself after the British system, in which there was both a public (BBC) and a commercial (ITV) station. However, in Finland, television merely had public and private "arms," called Yleisradio and Mainos-TV, respectively, rather than completely separate stations. Public broadcasting company Yleisradio produced commercial and free public programming, while industry and commercially owned Mainos-TV by leased broadcast time for its own production and resold it to private advertisers as commercial time for spots and sponsorships. The result was a hybrid system, in which some programming was purely public, while other programming was commercial. However, this system surely was not trouble-free.

term in 1968 which was used by both sociologists and left-wing youth politicians such as Paavo Lipponen (the editor of *Ylioppilaslehti* in the mid-1960s and Finnish Prime Minister 1995-2003)[14].

Repo's period in Yleisradio was stormy. Under his acceptance, particularly in television, young and radical programme makers and journalists were able to realise their programme which at times was strongly critical towards society and attacked bourgeious values, the church and patriotism. As a counterbalance to commercial bourgeous culture originating from America, Yleisradio aired very opinionated and, from the point of view of the right, very left-wing programmes. Due to this, the right-wing press, in particular, attacked *Reporadio*[15]. A stereotypical image of a so-called bearded radical – a young radical journalist with a beard and long hair – was born.

Reporadio was, in many ways, a product of its time. As cultural historian Marja Tuominen (1991, p.254) who has studied Finnish '60s generation in Mannheimian approach writes, at the end of the 1960s *Reporadio* was the most visible and loud symptom of the way that mainstream Finnish culture reacted, yielding to pressures demanding change in mentalities: "It was a fine example of the ability of Finnish society to simultaneously accept and be horrified, to channel, integrate and suffocate dynamics risen as counterbalance to the post-war restoration of values." The leftism of Yleisradio was often connected with the previously mentioned rise of the radical left taistoism at the turn of the 1970s as there were many taistoists among journalists.

On the other hand, the new radical generation also appeared in television commercials which it was criticising. In Unilever's Sunlight dishwashing liquid commercials, for example, the main character was an urban, educated baby boomer husband "Tiskaaja-Jussi" (John the Dishwasher), who did all the dishes in the household which he shared with his wife Ulla. This campaign began in 1968 and ran until the mid-1970s. Until 1972 the pattern of Sunlight commercials was pretty much the same: the man does the dishes and talks with his wife (who is always off camera). Usually the wife praises her husband for how good a husband and dishwasher he is. In one spot she even gives flowers to Jussi when he is celebrating a new, bigger Sunlight bottle and his 'everyday life'. But

---

14 The battle between commercial and publicly-funded television model also concerned other countries at times. For example, President of France Charles de Gaulle allowed commercials in the other channel of ORTF, the broadcasting company of France, as a kind of revenge both to television and press which had taken part in the strikes of the spring of 1968 (Kurlansky 2004, p.235).

15 As such, Repo's "informative and intellectually active programme policy" did not seem to be remembered. In my study on social history of Finnish television (Kortti, 2007), which material was people's TV-reminiscences, there are only a couple of mentions about it. Typically Reporadio was criticised by conservative and former elementary school teachers.

she can also be commanding. In another spot she says: "What are you mumbling about? You promised to do the dishes." You could say that Jussi is tied to his wife's apron strings.

Both actors were famous Finnish actors of the sixties. Heikki Kinnunen, who played Tiskaaja-Jussi, had a role, for example, in perhaps the best known Finnish play of the 1960s, *Lapualaisooppera*, which was very critical and radical – very "sixties". Kirsti Wallasvaara, who played his wife, Ulla, had one of the main roles in the very "European" new wave movie *Skin, Skin* (*Käpy selän alla* 1966), which was a story about students on their summer trip. It was the most significant Finnish movie of the 1960s. To cast commercials with such actors was a way to attach the new values of the sixties generation to the liquid. The couple stands for the new world view of the educated (in one spot Tiskaaja-Jussi is called MA), the emancipated sixties generation which mystically made Sunlight the dishwashing liquid of modern times. With the Tiskaaja-Jussi campaign the advertiser wanted to code new ideas about gender roles in order to differentiate from other dishwashing liquids, which basically were all the same. However, it was – cynically said in the marketing language – the advertiser's way to win a U.S.P. (Unique Selling Proposition) and overtake its competitors.

Radicals with their beards and long hair appeared more in commercials later in the 1970s. In this way, television was not only a new channel for the generation of baby boomers but it also commercially abused the image it had created of itself and its new modernist ideas.

8 Forum for Radicalism

However, it is important to notice that television did not rule communications even in its golden era in the 1960s, just like the Internet does not rule it nowadays. Of the electronic media, radio also had a significant role in the events of the year of 1968. During the occupation of the Sorbonne University in Paris in May, for example, the independent Radio One and Radio Luxemburg were significant channels for demonstrators. They brought the battles of barricades into the homes of Parisians. Radio journalists, who were regarded as following the events energetically and fairly, were treated as heroes among the occupiers (Caute, 1988, p.194). The American propaganda radio Radio Free Europa also had an important role in the movements in Poland (Kurlansky, 2004, p.71).

At the same time as television spread, the public sphere not delivered through the media was also blossoming in the form of demonstrations, teach-ins and other discussion events, for example. Often the only media devices in these public spheres were megaphones. Significant communications, particularly in the events of Czechoslovakia and France, also included writings on walls.

It is part of the present-day criticism of neo-liberalism to claim that the media is centralised and has become entertaining and commercialised which is also why the public sphere and the media's democracy has disintegrated in tje 2000s. However, the hegemony of Springer press in the Federal Republic of Germany in the 1960s, for example, meant that the concern ruled alone over 40 per cent of all the press of the Federal Republic of Germany and in West-Berlin even 100 per cent at times. The problems of centralised media, such as in Berlusconi's in Italy, are definitely not an issue of the 21st century of Western Europe[16].

For German generation leader of the 1960s, Ulrike Meinhof, for example, who in 1968 was one of the leading left-wing journalists in Germany and already a public figure, the bourgeois hegemony of the Springer press was part of her formative experience phase in life. After that Meinhof took the road of violence and terror as the German new leftism radicalised, which created the Red Army Faction, the Baader-Meinhof terrorist group, among others? Meinhof operated from the turn of the 1960s to 1969 in the student paper *konkret* – also as chief editor almost right from the beginning. Meinhof changed the paper from an anti-nuclear propaganda publication to a more general new-leftism opinion paper. Like *Ylioppilaslehti* in Finland, its circulation was by that time national and the position of *konkret* in the German public sphere was quite remarkable. When funding of the underground communist KPD party (Kommunistische Partei Deutchlands) was cut off in 1964, the paper became more commercial and magazine-like but its highly radical nature remained. However, Meinhof left *konkret* at the end of the 1960s and her philosophy of life turned to radical extreme and terrorism.

Student papers also had their role in radicalism in Finland where 1968 is a milestone. Chief editors in many Finnish student papers were fired due to too excessive leftism. The key event of the Finnish student radicalism in 1968 was the occupation of the Old Student House in Helsinki and one of the reasons for that was the policy of *Ylioppilaslehti*, the official publication of the Student Union of the University of Helsinki (HYY). After the post-war emphasis on culture, *Ylioppilaslehti* had again attained a radical label in the 1960s – however, now it was left-wing. It was regarded as a topical discussion forum – one of the most important one's in the country. Whereas the subtitle of *konkret* at the beginning of the 1960s was "magazine for culture and politics" (unabhängige zeitschrift für kultur und politik), in *Ylioppilaslehti* it was: "academic cultural political weekly".

Particularly during Jorma Cantell's chief editor period (1963-66), the policy of *Ylioppilaslehti* often collided with the views that its publisher HYY repre-

---

16 The media's strong commercialisation and centralisation already happened at the end of the 19th century during the reigns of the Anglo-American press emperors (Joseph Pulitzer, William Randolph Hearst and Lord Northcliffe).

sented. To calm the situation, the next chief editor to be appointed was Yrjö Larmola from the conservative National Coalition Party. As *Ylioppilaslehti* journalist of 1968 and visionary of Finnish new leftism Antti Kuusi (1968, 98) criticised his employer for being a deserter in his pamphlet *Ylioppilaiden vallankumous* (Revolution of students): "*Ylioppilaslehti* has managed to very skilfully cut out ideological discussion which a few years ago raged uncontrollably. It has indeed required plenty of phlegmatism, negligence and conservatism."

Ylioppilaslehti was not considered to fulfil the requirements set for it. This was evident particularly in how *Ylioppilaslehti* "left out" the events of 1968 in France, even though they were "the most significant ones during the history of the international student movement". Chief Editor Larmola had, according to his criticisers, printed "the paper, which had an essential position nearly as the only discussion forum in the whole of Finland, into mud". In fact, Larmola's policy was fairly moderate and space was also given to radicals. In *Ylioppilaslehti* of 1968 international student riots were also covered as well as other current events and phenomena. However, Larmola was too bourgeois for the Finnish radical elite of the student generation. He, among other things, judged demonstrating as "aping of foreign countries". Later in the 1970s *Ylioppilaslehti* again became an important forum for leftism and particularly for the Stalinist taistoist movement bred from it. At that time though, according to the new kind of democratic division political culture, the centre bourgeios tendencies were also given more space. At the same time, *Ylioppilaslehti* began to slowly lose its position as the voice of the generation.

The radicalism of baby boomers found other forums in the 1960s. Even though small free journals blossomed only ten years later in the punk movement, there were already different political and radical cultural papers in Finland at the end of the 1960s. The American alternative public sphere, at its most dynamic and in its most unruly phase, found its way to Finland in the form of an underground movement. The influences of the comic press – the American "comix conspiracy" – spread rapidly to Europe and also appeared in Finland particularly in 1968[17]. There was even an actual underground printing and publishing house and the freedom of speech association Demo operating in Helsinki.

The globality of the anti- and alternative public sphere was not created by the Internet. The ideas of the New Left generational movement were already

---

17 The alternative press agency UPS (The Underground Press Syndicate) was found in the United States in 1967 by a few underground newspaper and magazine publishers. They included, among others, *East Village Other*, which was assisted by legendary counter-culture figures such as Rober Crumb and Allen Ginsberg. According to certain "pre-internet ethics", all magazines which were part of the UPS were free to republish articles by others. The American underground press had hundreds of thousands of readers in the year of 1968 (Gitlin, 1980, p.164).

spreading internationally and fairly efficiently through publications of different countries in the 1960s. For example, one of the most distinguished periodicals of the movement was the British *New Left Review* which published articles by French and American thinkers at the end of the 1960s (Nehring, 2008, pp.131-132). On the other hand, the Italian periodical of the student movement *Quaderni Piacentini* co-operated with periodicals such as the *Studies on the Left*, *Revue Internationale du Sosialiseme* and *Monthly Review*. The last-mentioned was published in Italian from 1968 onwards. As Jan Kurz and Marika Tolomelli write in their account of the Italian protest and activism in 1968, these periodicals "served as hotbeds for international debate" (Kurz & Tolomelli, 2008, pp.86-87, 91).

Chief Editor of *New Left Review* Perry Anderson also visited Finland. Theorising of perhaps the most important thinker of the New Left and the student revolution prophet Herbert Marcus and colonisation philosopher Frantz Franon were also published in Finland in the year of 1968. They, like the rest of the media, were important in forming the world view for the elite generations of the 60s.

9 Conclusion: Discursive Dimensions in Media History

Finnish sociologist Semi Purhonen (2002) has criticised how academics who conceptualise generations, for example those with a Mannheimian approach, have not taken into consideration the "discursive dimensions" of generations; how a generation is produced in speeches, publications, manifests, et cetera. In other words, by paying attention to a discursive dimension, more specifically to the media, it can be seen that societal generations in a sense produce themselves (and others) discursively. For example, discourse concerning the 60s generation helps to create (to people of a certain age) common generation experiences – in other words, through discourse people begin to share common experiences more and more.

The academic youth has been very reflective. In other words, student generations have been consciously active to create their own generational identity. This appeared, for example in *Ylioppilaslehti* right from the beginning not only as concrete talking about one's own generation but also as reviews going backwards and forwards in the pages of the magazine and particularly in connection with the years which were marked by important events. As Finnish scholars of political history Timo Soikkanen and Vesa Vares (1998, p.38) state: "It is actually most important for a historian that the objects themselves confess the generation framework and the formative experience of the generation and behave according to it." Or as Robert Wohl (1979, p.5) states: "Historical generations are not born; they are made. They are a device by which people conceptualise society and seek to transform it." In Europe, at the beginning of the 20th century in particular, those who consciously called themselves the generation – generationalists, as Wohl refers to them – were very conscious of their own uniqueness and proud of their intellectual superiority. The means for conceptualising generations and changing society are often found through the media.

The media's role has also otherwise been particularly important and interesting in generational movements. Magazines and periodicals have often been a forum where discourse of a certain generation has been realised. The media's significance was emphasised in a new manner in the 1960s when television appeared, which was like the Internet of the 1960s. Naturally radio also had an important role, as well as even older media such as writings on the wall.

When a unit manages to create at least some kind of hegemonial discourse among coevals, we can talk about a generation. This happened among the Finnish academic youth both during the fascist-oriented inter-war period and in the 1960s and the 1970s of the New Left. One of the most important areas where hegemony was realised was their own medium, *Ylioppilaslehti*.

In order to analyse the media's role in generational movements it has to be placed in a broader historical context. It is media-historically important to collocate generational media in the spirit and mentality of the time: the media is powerful in creating Zeitgeist. However, if we want to do in-depth research into the media in generational movements we have to analyse how the philosophies of life which define the world view and activities of the whole movement are developed and realised through the media.

Because classical generation theorising is a modern creation, using it is often problematic when moving backwards from the pre-modern and particularly in the post-modern history. As Mannheim (1972, pp.105-120) states, all generational status, meaning the common features shared by individuals who are based on biological rhythm and sociological objective facts, do not always actualise as generation units. Currently the situation is that we live in a sea of rapidly flowing generation units. Since the 1980s and even before then, youth culture, for example, has been so fragmented that coherent generalisations often feel at the very least artificial.

Sociologists indeed say that "we" has atomised in the postmodern society[18] (see for example Lash, 1994, pp.143-156). Consumer researchers, on the other hand, see that consuming, consumer power and cultural taste are emphasised at the expense of communality in atomisation. (McKay, 1997, p.264) However, we

---

18 The concept of an atomised generation refers to a nuclear - or mosaic-like generation, the smallest parts that move vibrantly and dynamically in the field of cultural phenomena. Atomisation can be presented as a large scale process that cuts through a generation. The freedom and requirement for choice is characteristic for the life of the atomised generation. (Salasuo, 2006).

still have to remember that in connection with the use of the media, societies have not changed remarkably. There are still national states, strong collective identities and communities which extensively share the same interests (McQuail, 1997, p.145, 147), and I noticed this myself when researching the history of Finnish television viewing. In a sense, the identity of the "enlightened" or "sociological" subject has remained in television viewing and it has not changed to being very atomised (Kortti, 2007).

Additionally, it is essential to remember that in this era of blogs and other citizen journalism the media is still in the hands of the elite. Therefore, when we research media production, its views have to be considered separate from the whole mass, even though sometimes the views might meet.

Bibliography

Althusser, L., 2008. *On ideology*. London & New York: Verso.

Anderson, B., 1983. *Imagined communities: reflections on the origin and spread of nationalism*. London: Verso.

Burke, P., 1986. Strengths and weaknesses of the history of mentalities. *History of European Ideas*. 7 (5), pp.439-451.

Caute, D., 1988. *Sixty-Eight: the year of the barricades*. London: Hamish Hamilton.

Colombo, F., *The Long Wave of Generations*, Infra.

Davies, S., 2003. *Empiricism and history: theory and history*. New York: Palgrave Macmillan.

Gitlin, T., 1980. *The whole world is watching: mass media in the making and unmaking of the eft*. Berkley: University of California Press.

Gitlin, T., 1987. *The sixties: years of hope, days of rage*. Toronto, New York, London, Sydney, Auckland: Bantam Books.

Gramsci, A., 1996. *Prison notebooks*, vol 2 ed. Translated from Italian by Joseph A., Buttigieg. New York: Columbia University Press.

Kortti, J., 2007. *Näköradiosta digiboksiin: suomalaisen television sosiokulttuurinen historia*. Helsinki: Gaudeamus.

Kurlansky, M., 2004. *1968: the year that rocked the world*. New York: Ballantine Books.

Kurz, J. & Tolomerlli M., 2008. Italy. In: M. Klimke & J. Scharloth, ed. *1968 in Europe – a history of protest and activism, 1956–1977*. New York: Palgrave Macmillan Transnational History Series, pp.83-96.

Kuusi, A., 1968. *Ylioppilaitten vallankumous*. Helsinki: Otava. Delfiinikirjat.

Lash, S., 1994. Reflexivity and its doubles: structure, aesthetics, community. In: U. Beck, A. Giddens & S. Lash, ed. *Reflexive modernization: politics, tradition and aesthetics in the modern social order*. Cambridge: Polity Press, pp.110-118.

Manninen, J., 1977. Maailmankuvat maailman ja sen muutoksen heijastajina. In: M. Kuusi & R. Alapuro & M. Klinge, ed. *Maailmankuvan muutos tutkimuskohteena: näkökulmia teollistumisajan Suomeen*. Keuruu: Otava, pp.13-48.

Manninen, J., 1987. *Dialektiikan ydin*. Pohjoinen: Oulu.

Manninen, J., 1989a, Tiede, maailmankuva, kulttuuri. In: M. Envall & J. Manninen & S. Knuuttila, ed. *Maailmankuva kulttuurin kokonaisuudessa: aate- ja oppihistorian, kirjallisuustieteen ja kulttuuriantropologian näkökulmia*. Jyväskylä: Pohjoinen, pp.7-112.

Manninen, J., 1989b. Jälkiselvitys. In: M. Envall, J. Manninen & S. Knuuttila, ed. *Maailmankuva kulttuurin kokonaisuudessa: aate- ja oppihistorian, kirjallisuustieteen ja kulttuuriantropologian näkökulmia*. Jyväskylä: Pohjoinen, pp. 224–230.

Mannheim, K., 1972. The problem of generations. In: P.G. Altbach & R. S. Laufer, ed. *The new pilgrims: youth protest in transition*. New York: David McKay Company, Inc., pp. 101–137.

McQuail, D., 1997. *Audience analysis*. Thousand Oaks: Sage.

Mouffe, C., 1986. Hegemony and ideology in Gramschi. In: T. Bennet, G.Martin, C. Mercer & J. Woollacott, ed. *Culture, ideology and social process: a reader*. London: The Open University Press, pp. 219-234.

Nehring, H., 2008. Great Britain. In: M. Klimke & J.Scharloth, ed. *1968 in Europe – a history of protest and activism, 1956-1977*. New York: Palgrave Macmillan Transnational History Series, pp.125-136.

Purhonen, S., 2002. Sukupolvikäsitteen kolme ulottuvuutta: diskursiivisen dimension merkitys sukupolvitietoisuuden rakentumisessa. *Sosiologia* 1/2002, pp.4-17.

Purhonen, S., 2007. *Sukupolvien ongelma: tutkielma sukupolven käsitteestä, sukupolvitietoisuudesta ja suurista ikäluokista*. Research Reports No. 251, Department of Sociology, University of Helsinki.

Rintala, M., 1979. *The Constitution of Silence. Essays on Generation Themes.* Westport & London: Greenwood Press.

Salasuo, M., 2006. *Atomisoitunut sukupolvi: pääkaupunkiseudun nuorisokulttuurinen maisema ja nuorisotyön haasteita 2000-luvun alussa.* Helsinki: Helsingin Kaupungin Tietokeskus. Tutkimuksia 2006:6.

Sepp, H., 1997. Suomen silta kohtalon siltana 1919–1940. In: H. Roiko-Jokela & H. Seppänen, ed. *Etelän tien kulkija – Vilho Helanen (1899-1952).* Jyväskylä: Atena kustannus Oy, pp.39-88

T. Soikkanen & V. Vares., 1998. Sukupolvi selittäjänä Suomen historiassa. *Historiallinen Aikakauskirja.* 96 (1), pp. 37-53.

Stevenson N., 1995. *Understanding media cultures: social theory and mass communication.* London: Sage.

Suominen, T., 1997. *Ehkä teloitamme jonkun: opiskelijaradikalismi ja vallankumousfiktio 1960- ja 1970 -lukujen Suomessa, Norjassa ja Länsi-Saksassa* Helsinki: Tammi, Hanki ja jää.

Marja T., 1991. *"Me ollaan kaikki sotilaitten lapsia": sukupolvihegemonian kriisi 1960-luvun suomalaisessa kulttuuriss*a. Otava: Helsinki 1991.

Virtanen, M., 2001. *Fennomanian perilliset: poliittiset traditiot ja sukupolvien dynamiikka.* SKS: Helsinki.

Wohl, R., 1979. *The Generation of 1914.* Cambridge, Massachusetts: Harvard University Press.

Second Part: Generational Changes

Mariann Hardey

ICTs and Generations – Constantly Connected Social Lives

1 The First Generation of Social Media Users

The availability of ICT media has been associated with a 'privileged elite' and typically from middle-class households that had the financial means to be able to draw on a particular set of cultural resources to bring new technology into the home and to ensure access to the same resources at school (Becker, 2000; Holloway and Valentine, 2003; Livingstone, 1999). This relates to the broader debate about the nature of 'cyberspace' or a digital divide throughout the 1990s (Loader, 1998a; 1998b). In addition, class alignment and the age demographic of these young people have been used to characterise this group as 'native' or 'natural' technology users – a consequence of what Flacks (1971) identifies as the 'generational effects' of specific age cohorts.

At the time of writing this paper, this group of young people are members of one of the most recent and popularised generational classifications – a 'Generation Y' (e.g. Mitchell, 2003; Rugimbana, 2007; Ramsey et al, 2007; Dann, 2007). Typically Generation Y refers to the identification of a new and everyday style of mass-marketised and individualistic consumerism. The recent Mobile Youth Report (2008) suggested how the use of mobile media is upheld by the cultural status of technology in the eyes of younger consumers. For Facer, et al. (2001, p.451), Generation Y members correspond to the 1990s 'hype' about the 'myth of the cyberkid' and the groups natural interest and use of technology.

Other, perhaps more crude but useful, categorisations has been the emergence of such popular generation tagging as 'N-gen', 'Wired Generation', 'Echo Boomers', 'Google Generation', 'the iPods', 'iGeneration' and, one of the most recent and pertinent to this paper, the 'Facebook Generation' (Coupland, 2006; Hardey, 2007; Howe and Strauss 2000; Montgomery, 2007; Turrow and Tsui, 2008). These labels imply certain characteristics about the use of ICT media and characteristics or features of the demographic that they describe. Within this type of generational structuring, commentators have made reference to the potential for a 'generation gap' and considered how 'new generations' may be distinguished from previous generations (Pilcher, 1994, p.481). These popular names place emphasis on the stratification of populations that result from their anticipated technology competence and use. For example, in terms of the type of resources used, e.g. a 'Google Generation', and the expected engagement, knowledge and proficiency of the technology, e.g. as appropriate for a 'Wired Generation'.

2 Technology as a Cultural Status

When growing up in the 1980s with Sega Mega Drive, Nintendo and the first desktop PC's laid the foundations for a first generation of young technology users, the 1990s marked the convergence to more mobile media such as the mobile phone, cameras and MP3s that defined a 'way of life' and lifestyle. One of the most notable 'cult' products that has emerged as a prime identifier of youth in recent years has been Apple's iPod music player.

The first iPod was released in 2001 and marks a significant 'tipping point' in the consumerism and expressive identity of young consumers (Gladwell, 2002). As an example of a consumer product the iPod was deliberately marketed to have a lasting appeal to a 'young' and 'stylish' consumer demographic. Thus, the design was as much a part of the 'desirability' of the product as the function of playing and retrieving music files. More recently the iPhone has extended its appeal through a 'sexy' design and 'easy-to-use' user-interface, with a 'physically transparent' point of interaction that 'creates the feeling of a near-seamless interplay between the body and machine' (Beer & Gane, 2008, p.107). The iPod stands as a cultural symbol that had meaning not just peer-to-peer, but a social statement that could be interpreted by others, even if they did not share an interest in it. Gurwitch (2006) also makes the association between youth identity and the popularisation of what is a 'youth cult product', heralding the iPod as the 'electronic fountain of youth'. On the strength of this kind of youth consumption, American youth identity has been branded recently as endemic of an 'iPod Generation' (Katayama, 2007). The emergence of this particular youth culture is linked to consumer choice based on music taste, file shares and communities of music based 'friends' as part of a 'decentralised media', which has been recognised a part of the 'rise of a second media age' (Beer, 2006). Indeed, according to Bauman (1992, p.189) 'personal and selfdefinition', are exploited by the consumer culture that 'artificially' encourages certain '*styles* of life'. The new forms of communication developed quickly and became a part of the contemporary fashion and mode of social interaction, particularly amongst teenagers. Hence, the use of ICT media reflects and reinforces young people's emerging sense of identity and the ways in which they make sense of a complex social world. Moreover, the movement toward, what can be described as networked technologies can be understood as a shift toward a participatory and collaborative set of social and cultural engagement which hold particular appeal for young people especially when such engagement is focused around music

(Beer, 2008). The example of such technology includes the launch of the SNS MySpace in 2003.

3 The Emergence of Social Media and a Social Methodology

Web-based applications and technology have become increasingly pervasive and, as a consequence, significant for everyday social interaction. The research in this paper is based on ethnographic data from the United Kingdom and Australia that includes focus groups, interviews and participant observation methods. In total 160 participants contributed to the research situated at the universities of Melbourne and York, these included seven pilot interviews (see Table 1).

Interviews	Male	Female	Total
Pilots (University of York)	4	3	7
University of York	29	25	54
University of Melbourne	16	26	42
Totals	49	54	103

Focus Groups	Male	Female	Total
University of York	12	15	27 (6 focus groups)
University of Melbourne	13	17	30 (6 focus groups)
Totals	25	32	57

Figure 1: Tables showing the total number of participants included for the interviews and focus group data at the universities of Melbourne and York (Source: the author)

The data collection from October 2005 to September 2008 was designed to capture participants from both the initial wave of social technology use e.g. mobile phones and email, as well as the increasingly popularity of Social Network Sites (SNSs) including MySpace and Facebook. The main characteristics of these kinds of technology relate to openness, transparency and chiefly to *participation*. An important element to the research was that I as a researcher also occupied the same social context as the young people in the study and used the technology in the same way. In this way the research agenda shared the same social dynamics of the participants with particular emphasis on:

- Openness: The identification and inclusion of theoretical ideas that have emerged from the data, rather than 'squeezing' data into a predetermined theoretical frame, or selectively collecting data to fit a hypothesis (Strauss & Corbin, 1994);

- Transparency: The communication and details of the research exchanged with the participants;
- Participation: The recognition of my role as an ethnographer, which demanded a flexible and reflective analytical approach.

The emergence of what I label as 'social media' to include all web-based technology and the devices for accessing the internet and communication with others, has also been accompanied by shift in methodological approaches. Indeed much previous research that explored the Web 1.0 and 'cyberspace' focused on what Silver (2000) located as, 'cyber-cultural studies', virtual identities and communities that were only available on the web (Featherstone & Burrows, 1995; Rheingold, 1993). However, under Web 2.0, the social research possibilities and the consequences for academia remain relatively unexplored and relatively few studies have been focused on the social nature of Web 2.0 and consequences for doing research in this field (Beer & Burrows 2007; Hookway 2008). In this context, social media provides new opportunities and challenges for the observation and collection of research material, where a main component of the data may also include the research themselves (boyd, 2006; 2007[1]; Hookway, 2008).

4 Social Media and Everyday Social Practices

The first moment that the research participants began to utilise a range of social media for constantly connected communication was through networks built through SNSs such as Facebook and based on university membership. During 2005 after the initial launch of Facebook as an international platform there is a significant shift in daily co-present contact, from one that was based around a 'home life', family and friends, to the opening up of networks of friends as part of 'university life'. To draw attention to this transition, what follow are extracts from participants in December 2005.

Describing her move to university for Adele at the University of York this marked a period that was:

> 'frightening, and really intimidating. You're apprehensive about everything, from making new friends, to wondering what your old friends are up to, who you have stuff in common with, who you don't. Most of all I remember having moments when I just wanted to know what was going on. A few of my home friends had not gone to university, and so it was harder for them to appreciate all the changes, and why I didn't necessarily have time to see them anymore. They understood that I have moved away, but they also saw this as me moving on from them (...) looking back, it's still really emotional for me (...) I lost touch

---

[1] The author 'danah boyd' has changed her name to be published in lower-case text.

with some friends, not that we fell out, just that we were so used to seeing each other around, that when you are not around, you kind of fall out of sync. It's quite traumatic, and no-one prepares you for that! That's why I began to use Facebook. I was curious at first. Then I made up my mind to see if I could use it make better friends with other people who use it, and so it seemed the right thing to do'.

Adele felt *encouraged* to begin to use Facebook because of her university friends who shared the same social space. Her experience is typical of other participants in the study who acknowledged the same reasons for using social media such as Facebook. A student at the University of Melbourne, Chris mentioned that he,

'wanted to keep up with my friends, you've seen it for yourself, everyone here is on Facebook!'.

Another student, Carson, identified how (like Adele) he had been motivated by personal apprehensions, which centred on,

'worrying about my girlfriend (Holly). She went to a different university and I don't like being away from her. We text, we talk, but that was never enough (…) she has a cousin in the States who was on Facebook, and Holly is at Oxford and started on Facebook early. So now we message on Facebook and have our mobiles always on too'.

The upheaval and disruption that was experienced during the relocation to university were accentuated by the expectations that students like Adele and Carson already had about 'moving away', and chiefly, 'making friends'. Social media was used to counteract concerns about feeling 'frightened' and/or 'apprehensive', as well as curiosity about 'what friends were up to'. For these reasons, Facebook in particular held appeal with the research participants where it was seen to offer a 'stable' backdrop for social interactions during a period when social networks were seen to be shifting, in a state of flux and could be considered 'under threat'.

Where social media has begun to underwrite the everyday habits, social routines and friendships of the participants is typified by need to 'always be in touch' and convergence of multiple social networks that include friends 'at home', 'at university' and 'family'. For example, one of the main reasons Facebook had appeal amongst the participants was the initial loss of the routine contact with friends that was harder to reproduce at university. Friends were categorised in two ways, those who were 'at home' and, in effect seen to be 'left behind' and those who represented 'new friends' 'at university' who shared the same physical and social space(s). This divide was most prevalent amongst those students who joined Facebook *before* September 2006 when you had to have a university email address to join the site. Such exclusions marked 'active' and 'inactive' social networks where the site Facebook allowed for a means to potentially stabilise and sustain connections with others. At this point (before September 2006) there was a period of social adjustment marked by the first instance students around the world could join Facebook.

In Australia, there was a subtle, but notable difference in the way students first felt encouraged to join Facebook and described using the site. Generally, the students at the University of Melbourne were quicker to sign up to Facebook compared to those at the University of York, before September 2006. This was despite the timing of Facebook's platform launch in Australia in December 2005, some two months after the equivalent launch in the United Kingdom. One possible explanation was that in Australia it was more common for students to stay in their 'home' city, with established friends and family links closer to hand. Another difference was the variation in the provision of mobile telecommunications between the United Kingdom and Australia. Indeed, my own experience of purchasing and setting up a new Pay-As-You-Go mobile account in Australia, drew attention to the monopoly hold of the telecommunications company Telstra. This resulted in a limited choice of handsets, tariff rates, and network services. Thus, for the students at the University of Melbourne, mobile telecommunications was one of their main concerns, or 'gripes' on the university campus. This issue was repeatedly identifiable from the fieldwork data. For example, Rachel described how she was,

> 'dubious at first about Facebook (...) but then it was huge! There was this tidal wave of friends that were suddenly on it and all at the same time! I've used MySpace before, but everyone's so young and there's pressure to join groups that you don't really want to, and to friend just anyone and everyone (...) Facebook took hold cos it's just for students, and that was my friends and me (...) and I never get a decent signal on my mobile on campus, and its too expensive to keep calling everyone by chance, so Facebook's the best way to stay in touch'.

Rachel was not the only interviewee who felt that neither her mobile phone nor email provided a satisfactory contact with her peers. Tom, a friend of Rachel's, shared a similar motivation when he felt encouraged to join Facebook. Tom described his friends *'incessant use'* of Facebook to counter the 'crap network coverage on campus'. Increasingly there was concern was to know what was 'going on' with friends, as swiftly and as reliably as possible. Thus, connections began to rely upon the exchange of messages that could be 'picked up, read, and responded to' in a consistent and dependable form, with emphasis to be seen to act without delay.

At the University of Melbourne, as members of the university network, it was also usual for students to join one of the twelve associated college groups. The college groups were seen to reflect and endorse other student associations that had already been established at the university. For example, Chris explained when I first arrived to stay at Ormond College at the University of Melbourne, 'so you'll want to join the Ormond Group on Facebook. We're all on there, and that's where everything happens (…) it's the first thing I did before I knew anyone at Ormond, and the easiest way to make friends'.

By comparison, at the University of York, the college association was not as influential, nor seen as a significant element to identify with on Facebook. Students here made no mention of college groups, or any other shared associations on the site. By contrast, for Chris and other Melbourne students it was important that his initial encounters were encouraged through the collective association of a university college group. Chris described how he saw this as a way for his peers to,

'get to know each other, especially if they're shy (…) and it gives a feeling of solidarity'.

Hence, the students at the University of Melbourne were assured a familiarised context for interactions. Here the initiation of friendships was underpinned by college membership and further specialisation of the main university network. Clubs and societies that were also specific to a college were also coordinated across Facebook as well as the use of other social media such as Twitter. I spoke to a group of students from Trinity College, and Arnos mentioned,

'Our group is the best on Facebook! Go Trinity! (laughing). We're already organising the big row event for the row team on Facebook, that'll mean that everyone will be involved and can follow us on Twitter. It's easier than posting stuff in the Common Room'.

Interestingly in my first focus group, Adam dismissed Facebook university groups as, '*sad*', and later went on to pronounce that,

'Only saddo's join groups! What's wrong with your friends? Nah I'm already on a network, why would I want to join a group. The colleges are crap anyway!'

Rather than a college-based set of associations, of more central importance to students at the University of York was the accumulation of links that were formed around the main university network. College groups were viewed as 'unnecessary', and 'uncool', and hence had very limited appeal. At the time of the focus group that included Adam, held January 2006, only three of the eight colleges at the University of York had affiliated groups on Facebook. We can speculate the reason for the difference in interest and membership of college groups. This could be to do with the comparative size of the student population at each university. Both the campus size and college population at the University of Melbourne is significantly greater than that at the University of York. As a guideline, the headcount for the number of undergraduates in 2006 at the University of York, was 7,762, and at the University of Melbourne, 28,843. It was noticeable that at the university of Melbourne the students who had been encouraged by their college association formed stronger community attachment to others as a result of their group and college membership. One participant mentioned how the college group he was a member of, 'feels like a real community away from home'. This was quickly matched by participation in other events and activities within the college, 'on the college page you always know what's going on, that's enough reason for me to use Facebook'. Indeed, whilst students at the University of York are formally arranged around college membership, there is less emphasis on these associations as an integral part of the students' experience.

At both the Universities of Melbourne and York, Facebook took hold, because the students had felt encouraged, and began to collaborate with friends who were already using the site. The boundaries between the previously online and the offline communication of email and IM, and the 'on' and 'off' of mobile phone coverage, declined in importance, particularly with the convergence of several social media resources that typically included Facebook, Twitter and the ownership of at least one mobile phone and access to the internet. It was significant that connections were built up and held in place by the *known* connections of friends and the membership of relevant networks as a university, and/or college group/association. Together these situate such social practices as everyday and essential interactions for near-instant communication.

5 Conclusion: New Social Practices and Etiquette

From 2005 there were increasing numbers of students who quickly adopted Facebook as a new space in which to conduct their social lives. The migration to Facebook relates to the way that various users have noted that you were somehow 'left out', if you did not use the site or have access to the same social information. Facebook, while important for establishing and shaping friendship and social activities amongst university students, also became used as a way to 'catch-up', or included friends and others who were left at home. Indeed, once students were set-up on Facebook, there is a sense of how their connections are consistent with already established friendship dynamics and therefore quickly become embedded within everyday sociability. These observations build on the desire to be 'one of the many' with the incorporation of social interaction into daily life that can be 'taken for granted' and represent a set of new social practices and etiquette. Thus, it is necessary for the individual to have access to numerous social media that act as a platform for all interactions. How these various elements are cast are designed to converge seamlessly, and to relate to consistent and detailed sets of social information that is constantly broadcast to friends. Here the emphasis is on the *continuity* of the expression of social action which is reproduced as part of a reflexive engagement with known others. As a result, pre-established social norms and values are reproduced across new social spaces and become established as new sets of social practices and the display of a particular social etiquette. In this context it is paramount that friends can 'recognise' each other both 'in-person' and through various social media. Here there is a growing awareness of about the participation, visibility and obligations with others as users are concerned to 'feel secure' within social networks. Thus the research participants can be viewed as 'modern subjects' or 'actors', where the opportunity to construct a 'new' self, and to celebrate 'new' relationships is safeguarded by reproduction and replication of known social processes.

This paper is based on the relatively 'small-scale' and intimate interactions between 'alike' others to suggest a set of nuanced social techniques in the management of social connections (Hardey, 2008a; 2008b). Such techniques can be read as a form of social etiquette that is based on the self who is *always* connected *with* others. This significance has been captured in my blog 'Proper Facebook Etiquette[2]' that explores the everyday and shared dilemmas as a result of interactions managed by and pursued through social media. Such strategies highlight the capacity for what we can begin to speculate as to the 'opening up' of social dimensions, whether through Facebook, Twitter etc. as well as a 'closing down' and/or control of shared social information. Indeed, some of the more critical commentaries have been about the harmful and potentially isolating effects of the technology (e.g. 'Facebook Child Porn Case Results in 35-year Federal Prison Term[3]', *Chicago Tribune* (2008); 'Facebook Addiction Can Kill[4]', *The Times* (2008); 'Man Killed Wife in Facebook Row[5]', *BBC News* (2008).

For the participants in this study as part of a particular generation of young users there is a sense of a growing awareness of the possibilities of the technology where such connectivity is 'taken for granted' and revealed in a mostly positive manner. Where such responsiveness is shared across a generation led by the most enthusiastic users there are important consequences about how to manage social information and the long-tail effects of data 'ever-present' in numerous 'public' domains. Such social practices I argue are seen to reinforce, and occasionally compete, with understandings for successful interaction. Above all, the understandings of such sociability represent essential elements to the recreation and translation of social actions across numerous social media. Together these are suggestive of a particular sociability that represents changes

---

2   http://properfacebooketiquette.blogspot.com/
3   http://www.chicagotribune.com/news/local/chi-fox-lake-facebook-both-22oct
    22,0,2137581.story
4   http://www.thetimes.co.za/News/Article.aspx?id=867888
5   http://news.bbc.co.uk/1/hi/england/london/7676285.stm

appropriate to, not only this generation of young people, but increasingly their friends and friends-of-friends as an ever-increasing network of links.

Bibliography

Bauman, Z., 1992. *Intimations of Postmodernity*, London: Routledge.

Becker H. J., 2000. Who's Wired and Who's Not: Children's Access to and Use of Computer Technology, *Future of Children Organisation*, [Online], Available at: http://www.futureofchildren.org [Accessed October 2008].

Beer, D & Gane, N., 2004. Back to the Future of Social Theory: An Interview with Nicholas Gane, *Sociological Research Online*, [Online] 9 (4), Available at: http://www.socresonline.org.uk/9/4/beer.html [Accessed June 2006].

Beer, D., 2006. The Pop-Pickers have Picked Decentralised Media: the Fall of Top of the Pops and the Rise of the Second Media Age', *Sociological Research Online*, [Online]. 11(3), Available at: http://www.socresonline.org.uk/11/3/beer.html [Accessed October 2008].

Beer, D., 2008. The Iconic Interface and the Veneer of Simplicity:Mp3 Players and the Reconfiguration of Music Collecting and Reproduction Practices In The Digital Age, *Information, Communication & Society*, 11(1), pp. 71–88.

Beer, D., Can you dig it? Some Reflections on the Sociological Problems Associated with Being Uncool, *Sociology.* Forthcoming

Beer, D. & Burrows, R., 2007. Sociology and, of and in Web 2.0: Some Initial Considerations, *Sociological Research online*, [Online]. 12(5), Available at: http://www.socresonline.org.uk/12/5/17.html [Accessed June 2008].

boyd, d.m., 2006. Identity Production in a Networked Culture: Why Youth Heart MySpace, American Association for the Advancement of Science, [Online], Available at: http://www.danah.org/papers/AAAS2006.html [Accessed August 2008].

boyd, d.m., 2007. Why Youth (Heart) Social Network sites: the Role of Networked Publics in Teenage Social Life, in D. Buckingham ed. MacArthur Foundation Series on Digital Learning Youth, Identity, and Digital Media Volume, Cambridge, MA: MIT Press.

Coupland, D., 2006. *JPod*, London: Bloomsbury.

Dann, S., 2007. Branded Generations: Baby Boomers Moving into the Seniors Market, *Journal of Product & Brand Management*, 16(6).

Facer, K. Furlong, J., Furlong, R. & Sutherland, R., 2001. Constructing the child computer user: from public policy to private practices'. *British Journal of Sociology of Education*, 22(1).

Featherstone, M. & Burrows, R. ed., 1995. *Cyberspace/Cyberbodies/Cyberpunk: Cultures of Technological Embodiment*, London: Sage.

Flacks, R., 1971. *Youth and Social Change*, U.S.A: Rand McNally & Co.

Gladwell, M., 2002. *The Tipping Point,* USA: Back Bay Books.

Gurwitch, A., 2006. The iPod, an Electronic Fountain of Youth . *NPR Pop Culture*. [Online]. Available at: http://www.npr.org/ [Accessed November 2007].

Hardey, M., 2007. Converging Mobile Technology and the Sociability of the iGeneration. *M/C Journal*, March 10.

Hardey, M., 2008a. The Formation of Rules for Digital Interaction, *Information, Communication and Society*.

Hardey, M., 2008b. Seriously Social: Making Connections in the Information Age, Ph.D. York: University of York, Department of Sociology and Centre for Women's Studies.

Hardey, M. (continuous) *Practising a Proper Social Demeanour: A Guide To Facebook Etiquette*, [Online]. Available at: http://properfacebooketiquette.blogspot.com/ [Accessed November 2007].

Holloway, S.L. & Valentine, G., 2003. *Cyberkids: Children in the Information Age,* London: Routledge Falmer.

Hookway, N., 2008. Entering the Blogosphere': some strategies for Using Blogs in Social Research. *Qualitative Research*, 8(1), pp.91-113.

Howe, N. & Strauss, W., 2000. *Millenials Rising: The Next Great Generation*, New York: Vintage Books.

Katayama, F., 2007. American Youth: Ipod Generation', *Reuters UK*, [Online]. Available at: http://uk.reuters.com/ [Accessed November 2007].

Livingstone, S.M., 1999. Personal Computers in the Home: What Do They Mean for Europe's Children? *Intermedia*, 27(2), pp. 4-6.

Loader, B.D.,1998a. Cyberspace divide: Equality, Agency and Policy in the Information Society, in B. Loader ed. *Cyberspace Divide: Equality, Agency And Policy In The Information Society*, London: Routledge.

Loader, B.D., 1998b. Equality, agency and policy in the information society' in *Cyberspace Divide, London*: Routledge.

Mitchell, W., 2003. *Me ++,* MIT: MIT Press.

Montgomery, K.C., 2007. *Generation Digital Politics, Commerce, and Childhood in the Age of the Internet,* MIT: MIT Press.

Pilcher, J., 1994. Mannheim's Sociology of Generations: an Undervalued Legacy', *British Journal of Sociology,* vol.45, pp.481-494.

Ramsey, R. P. Marshall, G. W., Johnston, M. W. & Deeter-Schmelz, D. R., 2007. Ethical Ideologies and Older Consumer Perceptions of Unethical Sales Tactics, *Journal of Business Ethics,* 70(2).

Rheingold, H., 1993. *The Virtual Community,* Reading, Mass: Addison-Wesley.

Rugimbana, R., 2007. Generation Y: How Cultural Values Can Be Used to Predict their Choice of Electronic Financial Services, *Journal of Financial Services Marketing,* 11(4).

Strauss, A. & Corbin, J., 1994.Grounded Theory Methodology; an overview, in N. K. Denzin and Y. S. *Lincoln Handbook of Qualitative Research,* London. Sage.

Turow, J. & Tsui, L. ed., 2008. *The Hyperlinked Society Questioning Connections in the Digital Age,* The University of Michigan, The University of Michigan Press.

Giovanni Boccia Artieri

Generational "We Sense" in the Networked Space. User Generated Representation of the Youngest Generation

1 Media and Generations: Theoretical Perspective

The main goal of the research project we are discussing here[1] is to understand if and how media products (e.g. novels, movies, TV shows) affect the wide process leading to the creation of a shared set of meanings and sense of belonging, what Bude (1997) called generational 'we sense'.

According to Mannheim's classical definition of the sociological concept of generations (1952), a generation arises when the youth experience the same concrete historical problems. This classical definition, which in a direct way links the construction of a set of shared meanings with historical problems, even if it provides a fruitful starting point for every further analysis, it doesn't explain much about how generations arise. It has been argued that people who grew up together and faced the same historical problems can be as well defined by using the "technical" concept of 'cohort' (Ryder, 1965; Hine, 2000).

The cohort concept keeps together people according to their birth year and with regard to the characteristics they share (Corsten, 1999). Compared to the concept of generation, the cohort one seems to be much more easy to define and it allows a large comparison between different birth cohorts. In spite of the success of the cohort concept the generational concept seems still to be attractive for sociologists.

The last theorizations about generations try to explain the concept not as a mere construction operated by the sociologist (Corsten, 1999; Edmunds, 2002), but as a multi-dimensional issue which assumes a shared assumption of a common life experience by the members and the emergence of that specific generation as a social-fact itself.

The generation, as a theoretical concept, then assumes a reflexivity process (Giddens, 1991) both at an individual and at a collective level (Bech, et al., 1994). Recently Edmunds and Turner (2005). Proposed that mass media might be a common landscape able to offer a world-wide shared stage for "concrete

---

[1] This paper is an excerpt of the research PRIN 2006 "Media semantic and Generations: cultural experiences and collective identities in non-urban scenario". Researchers Giovanni Boccia Artieri, Laura Gemini, Fabio Giglietto, Luca Rossi – LaRiCA, University of Urbino "Carlo Bo". Available at: http://mediageneration.wordpress.com/.

historical problems" experienced by a growing *global generation*. The global *mediascape* (Appadurai, 1996) offers the possibility for young people all over the word to experience, for the first time in history, a global view of the world and, as a consequence, a global dimension of problems.

From the Vietnam war to 9/11 several waves of global, media-based, generations might experience specific localised problems as world-wide issues (Boccia Artieri, 2004). Parallel to this phenomenon the growing of digital practices related to the User Generated Content offers to the audiences an increasing number of way to self-observe themselves and to represents the others (Boccia Artieri, 2006, 2008).

If the generational 'we sense' may be described from a more hermeneutical and linguistic point of view as a meaningful set of connected criteria for interpreting and articulating topics in communication (Corsten, 1999; Luhmann, 1980), then mass media would be the place where those 'criteria' are learnt. Edmunds and Turner's assumption about a global generation (Edmunds & Turner, 2005) moves a step forward the whole generational research. If, as we claimed before, generations are a shared construction that cannot exist without a specific generational perception between the members, how can this "we sense" in world-wide media based generations be observed?

Weblogs, and the whole web 2.0, seems to be a viable place where to observe if and how generational discourses emerge, and if they may arise around specific media-contents or media-topic able to trigger the reflexive process.

The reality of the web 2.0 represents a networked space where the generational "we sense" can be produced and observed. In social networks sites, inside the conversational reality of blogs, besides the video of everyday life posted on YouTube we can observe ourselves while we're telling our story and we can see the stories told by people that we "feel" like us. The experience is networked trough the friendship languages of Facebook or trough the references system of the blogosphere: the reflexivity and the networked practices come into resonance in the reality of world wide web. Here even the forms of hetero-representation generated by media products-movie, tv fiction, anime, etc. - become an opportunity for self representation providing us the tools and the raw material for the the production of new meaning in the UGC form: remix, mash up, etc.

The media products are, in this sense, a background where generations and sharing of generational narratives made of conversations around them became possible. These networked practices (re-production, sharing, conversations) trigger mechanisms of reflexivity that link an individual with a collective reality, i.e. produce a "script" for the generational wesense.[2]

---

2 For general reflections on generational we-sense and the relation with media see Colombo, infra.

Today, in our society, the digital natives (Palfray & Grasser, 2008) are producing their generational identity in a networked space between online and offline reality:
"From the perspective of a Digital Native, identity is not broken up into online and offline identities, or personal and social identities. Because these forms of identity exist simultaneously and are so closely linked to one another, Digital Natives almost never distinguish between the online and the offline versions of themselves.

Digital Natives establish and communicate their identities simultaneously in the physical world (the sixteenyear-old might be bound to being a tall Irish-American girl), and digital space (where she can experiment with self-representation, sometimes in modest ways and sometimes dramatically, and her multiple representations inform her overall identity)" (Palfray & Grasser, 2008, p.25).

To observe this online reality means to understand the ways in which youngest will produce their generational representation trough conversations and feelings about the cultural products they lives in – particularly, as we say, the media products.

2 Reflexivity in Networked Space

This paper will propose to investigate this specific and furtive phenomena in the so-called web2.0 that, for the first time in history, moves the audience to the producers' side (Jenkins, 2006a, 2006b; Boccia Artieri, 2008).

The emergence of the convergence culture (Jenkins, 2006a) between media producers and audiences gives to the audiences the ability to interact with mainstream media companies. This modified paradigm is forcing media studies to investigate, as a brand new research field, how grassroots participation can interact with institutional media companies. Small groups communications live today in the very same media environment in which occur broadcast communications, both are parts of an interconnected ecosystem where UGC and institutional media production can resound together.

Contemporary audience studies have shifted their theoretical paradigms from the old conceptions of general audience to the idea of performing audiences which stress the active role of the spectator in understanding, negotiating and reshaping the meaning of media products. This scenario seems to evolve today toward a more complex one able to keep together (media-based) individual reflexivity processes with the perception of being digitally networked with others. This happens both in social network sites (boyd, 2008) and in networked publics experience (Kazys, 2008).

If, as we stated before, generational processes require a double level of reflexivity (both individual and generational) weblogs are the perfect place where to observe reflexive processes at large. In the blogosphere the distinction between authors of media contents and audience fades: this is the reality of a networked space. The author of a blog perceives him or herself by reading what he/she writes as a reader would do. Authors and readers, in turns, use their own life experience as a part of the communication process. And both start a reflexive process in order to understand life-experiences told by someone else. Those online conversation are a fruitful resource for every sociological investigation and some specific characteristic could offer a brand new and exciting scenario to the whole sociological research. Online digital user generated contents shows some proprieties: *persistency, searchability, replicability and scalability* (boyd, 2008). Each of these properties is nothing new in the media landscape but all of them have never been present in one media before the World Wide Web.

- Persistence is one aspect that online communications share with other media (eg. writings). It is possible to access online communication even after the communication event has taken place and sometimes link structures describe the process of reproduction of communication from communication. This point is useful for longitudinal research.
- A large amount of data is not always useful if there is not a way to filter and sort the data according to research questions. Online conversations as digital texts are easy to search (e.g. Google). In addition, digital photos and videos are often easy to search thanks to the tags that users use to apply to categorise their own contents. At the same time the structure of tags become observable items and can be even used as data (e.g. social tag tools such as deli.ci.ous).
- Online conversations are easy to replicate. This property increase the possibility of reuse and mix of available contents and create the possibilities to observe, along with the citation link structure, the process of reproduction of communications.
- The scalability defines the potential visibility of content in networked publics. On the Internet everyone can be an author: "Internet made it possible for anyone to broadcast content and create publics, although it did not guarantee an audience" (boyd, 2008, p.29).

The crossing of these four properties generates on one hand the possibility for social scientists to start using, as research data, the persistent user generated content and on the other hand the opportunity, for a wider audience, to have many more opportunities for reflexivity processes. From a methodological perspective UGCs can be considered as a new kind of social data. Since they are autonomously produced UGCs move a step away from the traditional qualitative data and the empirical consequences of the four properties listened before are the ability to retrieve automatically a large amount of data for research purposes.

If online discourses are autonomously produced and addressed to an unknown (and unknowable audience) they are incredibly rich for qualitative analysis since they potentially solve many of the distortions introduced by specific qualitative tools such as interview and focus-groups. At the same time working with searchable and persistent data makes much more easier for the researcher to avoid the community-bounding limits that can be saw in virtual ethnography practices. While virtual ethnography is heavily based on "the situated presence of an ethnographer in the field setting, combined with intensive engagement with everyday life of the inhabitants of the field site" (Hine, 2000) working with searchable and persistent data will enlarge research possibility by maintaining, at the same time, the qualitative value given by an autonomously produced content.

Can these online resources be used to investigate how people carry on generational discourses online? If, as claimed by Edmunds (Edmunds & Turner, 2005), media can be the shared stage where people can experience the same facts in spite of the geographical belonging, can weblogs be a viable place where to search for generational reflexivity?

According to the proposed theoretical and methodological references the paper will propose the result of a wide research that, using a specifically developed software tool (WeSearch), investigated more than 3000 Italian blogs in order to observe how reflexive processes were triggered by historical media products.

3 Wesearch Methodology

In order to investigate media-related generational discourses on the web we have developed a research process called 'WeSearch'. It has been articulated in four steps: a) identification of generational products, b) definition of query strings for online analysis, c) retrieval of blogs author's information d) content analysis of collected data.

3.1 Identification of Generational Products

The research process started by identifying a set of media products (movies, music albums, TV-shows, books and comics) which had been labeled as generational by their audience[3]. In order to do so we carried on 5 group interview with 3 participants belonging to generations X and Y, who were born between 1966 (the beginning of Generation X) and 1991 (the last birth year of generation Y). That specific generational time-frame had been chosen according to an existing literature (Aroldi & Colombo, 2007). Each group had been asked by the interviewer to identify a set of media products that they would consider "generational". During the interviews the researchers focused the attention not only on media products but also on how media had been experienced and how media-products entered the generational discourse. This phase ended with a set of 45 media products (chosen from the top 9 of each category).

3.2 Definition of Query Strings for Online Research

In order to collect all the Italian-language blog posts dealing with selected media products, every media-product had been used as a keyword to query Google Blog Search. Even if we were aware of some of the weakness of Google Blog Search for social analysis, we decided to use Google's service in order to collect blogs entries mainly for two reasons: 1) it returned less SPAM blog posts than similar services and 2) the language filter worked better than competitors such as Technorati, allowing us to focus the research only on Italian blogs.

Google blog search had been queried in order to obtain an RSS feed containing top 100 entries ordered by relevance. We opted for a relevance sorting because of the non chronological nature of collected data. The outcome of this phase was a standard RSS feed generated by Google's service, composed by every retrieved blog entry containing selected keywords. The first problem that the research team encountered was that sometimes blog services did not provide an RSS feed containing the full text of the blog entry. To solve this problem it was necessary to develop a tool able to fetch the full text content of every single article. We chose to use the 'Yahoo pipes' services in order to develop a small tool aimed to carry on this work for us. The final output of this phase is an RSS feed containing the full text of every single blog entry containing the selected keyword.

---

3   The research team did not provide a definition for the term "generational" but let the participants interpret the term according to their own perception.

3.3 Storage of Blog Entries and Retrieval of Authors' Information

The RSS feeds was stored in an incremental way using a software which was specifically envisioned by the research team. The web application is able to store an RSS feed, sorting the stored information by authors.

Another feature of this software is the ability to retrieve biographical information about entries authors. Obviously this information may be obtained only if the author him/herself added them to his/her own online profile. Biographical information (age, gender and location) is retrieved using a scraping technique specifically developed for every blogging or sharing service. Currently there are 7 supported platforms, as shown in table 1.

Service	Scraping languages	url	Unique audience [000][3]
Blogger	ITA / ENG	http://www.blogger.com	5190
Flickr	ENG	http://www.flickr.com	N/A
Il Cannocchiale	ENG	http://www.ilcannocchiale.it	590
Libero	ITA	http://www.libero.it	4643
Splinder	ITA	http://www.splinder.com	2303
Windows Live Space	ENG	http://space.live.com	5063
YouTube	ENG	http://www.youtube.com	N/A

Table 1: Service scraping languages url unique audience[4]

All the collected information (both the posts and the biographical data about authors) may, at any time, be exported in textual files (.doc format). The exported files will be compatible with the Nvivo7 qualitative analysis software. Biographical information about authors will be exported in a comma separated data file which can be easily imported as an Nvivo casebook retaining the link between the biographical information and the text entries.

3.4 Summary of Collected Data

Starting from the set of 45 media products used as starting keywords, the research was able to retrieve around 3000 blogs entries. Author's information was available for 928 cases (31%) of which 49.34% was male and 50.76%

---

[4] Data courtesy of Nielsen//Netratings Netview. Standard Metrics (Internet Applications Included) February, 2008, Country: Italy. YouTube and Flickr was not considered.

female. Generation Y is the most represented with the 79% of blogs entries[5], generation X scores the 15% and Boomer Generation (born between 1953 and 1965) and postwar (born between 1940 and 1952) score both 3%. The collected data were analyzed from the qualitative and quantitative point of view using Nvivo7 software.

Before stepping into the research results and move back to the relation between media products and generational we sense, it is now possible to make few preliminary considerations on the adopted methodology. One of the most interesting things we noticed is somewhat a confirmation of what it is already well known about American blogosphere. Italian bloggers use to share thoughts and personal information (e.g. age, gender on profile) online in a massive way. The scraping technique we developed (that can be easily extended thanks to its modular structure) was able to collect authors' information for more than 1/3 of the whole set of data (31%). The data confirm that the vast majority of active Internet users (blog authors, people who share pictures or video online, etc.) may be located in the youngest part of the population. According to the previously mentioned Research (LaRica, 2008) the 74.3% of Italian blogs authors is between 18 and 29 years old. Overall results are, of course, heavy influenced by this age distribution.

Together with those encouraging results we also experienced some limits of the methodology. The first obvious objections concern the author's biographical data in the profiles. We are well aware that data in user profiles may not be accurate since the users may decide intentionally to deceive while filling the registration form. At the same time we also know that both post contents and data in the profile are conversations and we therefore observe them as a social construction. On this level of observation what matters is the realness of the construction itself and not its relationship with reality (Luhmann, 1980).

4 Posting about Generational Identity[6]

With this research we're observing that generation that has been called Y generation. Since this is the last "completed" generation it is the only one we can observe. Here, in the differences with the previous X generation we can spot some traces of how the next one, the digital natives generation will use the networked context to show new media-based reflexivity processes.

---

5   Obviously all the information about generations is based on the smaller set of blogs entries with author's information available.
6   All the blog excerpts were translated from Italian to English by the researchers.

The huge number of retrieved posts pinpointed the capability of media products to trigger generational discourses. Generational discourses, triggered by the specific media product, seem to be used in two different ways. Sometimes the specific product gives the opportunity to start a reflexive process narrating single and personal events. Media product may then act as specific keys to release hidden memories or as pivotal elements in the personal biography.

> "I chose to start the nursery class. My nurse-syndrome and the fact that I watched *Candy Candy* when I was a child, made me choose this university degree" (F., female, 19 years old).

Media-products are perceived as something deeply related to the individual life and identity creation. Despite that, at the same time, media products seem able to start wider reflexivity processes. Those processes, which may be considered as a truly generational form of thought, seem to link the specific media product to a wider and shared *we sense*.

> "Today there are no more values in society… but today's television is partially guilty… do you remember the cartoons we used to watch and how they taught us to love and take care of feelings?! Candy Candy, Lady Oscar and Georgie… What today's cartoons are about is only violence!!!!" (F., female, 25 years old)

There is the feeling of something shared. Something that is assumed to be common because of the sharing of a specific media product or of a specific time in media history:

> "And we who are in the age of thirties, we belong to the Tiger Mask generation, we can't change it…We had good times, watching cartoons and TV series that left their mark." (A., male, 32 years old)

This kind of double reflexivity that media products allow (individual reflexivity and generational reflexivity) is strongly related to the Italian media history and to the heavy diffusion during the seventies and the eighties of media technologies. Media technologies, due to the high impact they have in everyday life-routine may be seen as milestones used to label specific time in personal and generational history. Generation definitions, when observed from the media perspective, seem to be a mixture of specific media-products and specific media technologies:

> "We the generation born in the seventies and grew up in the eighties… we're the last *naïve* generation. We didn't have super videogames and virtual reality, our games were the crazy ball, and the gluing hand… the bicycle (Graziella [a folding bicycle] for girls and BMX for boys). The top edge technology we had was Pac-Man" (E., male, 30 years old).

5 Conclusions

The good quantitative result of the analysis demonstrates that media-products are really good triggers for generational discourses. It is interesting that the specific-media product is not necessarily the topic of the produced content. People do not usually speak about that media-product but they use it as a specific mnemonic anchor. A single product may be used to help the reader remember what that time was or what the general "mood of the time was". Truly generational media-products do not require introductions or explanations since the writer assumes that readers know them. The mutual knowledge is taken for granted. F., in our first example, does not care to explain what the relationship is between Candy Candy and the nursery since she assumes that everybody should know it. If media-products seem to be very good anchors for memories, the generational use of these memories seems to be a little bit more complex. Authors usually talk about media contents both in an individual and in a generational way. We consider an individual use of the media-product when the product is linked to a specific event of the user's personal life. At the same time a truly generational use of the media-product may be observed when the product is used to evoke a shared knowledge or feeling. In this second case the media-product acts as a strongbox for a shared we sense which may be summoned by the media-product. In the reported example A. speaks about Tiger Mask in order to disclose a shared knowledge not related to the anime itself but about the generation of the viewers. It is like a secret phrase known only to some.

Bibliography

Appadurai, A., 1996. *Modernity at Large: Cultural Dimensions of Globalization*. Minneapolis: Public Worlds. University of Minnesota Press.

Aroldi, P. & Colombo F., 2007. Generational Belonging and Mediascape. *Journal of Social Science Education* 1.

Bech, U. Lash S. & Giddens A.,1994. *Reflexive Modernization: Politics, Tradition, and Aesthetics in the Modern Social Order.* Stanford California: Stanford University Press.

Boccia Artieri, G., 2004. *I Media-Mondo*, Meltemi, Roma .

Boccia Artieri G., 2006. Farsi Media. Consumo e Media-Mondo: Tra Identità, Esperienza e Forme Espressive. In Di Nallo E. & Paltrinieri R., *Cum Sumo. Prospettive Di Analisi Del Consumo Nella Società Globale*. Milano: Francoangeli.

Boccia Artieri G., 2008. *Share This! Le Culture Partecipative Nei Media. Una Introduzione A Henry Jenkins*, In: H., Jenkins, 2006. *Convergence Culture*, New York:University Press New.

boyd, d., 2007. Why Youth (Heart) Social Network Sites: The Role of Networked Publics. In D. Buckingham ed., *Teenage Social Life. Youth, Identity and Digital Media*. Cambridge MIT Press, MA, pp. 119-142.

boyd d. , 2008. *Taken Out of Context: American Teen Sociality in Networked Publics*. Ph. D., Berkeley: University of California- School of Information.

Bude, H., 1997. *Das Altern einer Generation. Die Jahrgange 1938-1948*. Frankfurt am Main: Suhrkamp.

Colombo, F., *The Long Wave of Generations*, *Infra*.

Corsten, M., 1999. The Time of Generations. *Time & Society*, 8(2), pp. 249-272.

Donati, P., 2007. Generations. A new research agenda in sociology of culture. In: Università Cattolica di Milano, March, 23th 2007.

Edmunds. J., 2002. *Generations, Culture And Society*. Philadelphia: Open University Press.

Edmunds J., Turner B., 2005. Global Generations: Social Change in the Twentieth Century, *The British Journal of Sociology*, 56 (4), pp. 559-577.

Giddens, A., 1991. *Modernity and Self-Identity: Self and Society in the Late Modern Age*. Stanford: Stanford University Press,.

Glenn, N. D., 1977. *Cohort Analysis*. Beverly Hills: SAGE.

Hine, C., 2000. *Virtual Ethnography*. London: Sage Publications Ltd.

Kazys V. (ed.), 2008. *Networked Publics*. Cambridge: MIT Press.

Jenkins H., 2006a. *Convergence Culture*, New York: New York University Press.

Jenkins H. , 2006b. *Fans, Bloggers, and Gamers. Exploring Partecipatory Culture.* , New York: New York University Press.

LaRiCA, 2008. I Social Media in Italia: oltre la Punta dell'Iceberg. Avalaible at: http://laricavirtual.soc.uniurb.it/?p=39 [Accessed March 2009].

Lenhart A., Madden M., 2007. *Teens, Privacy & Online Social Networks*. Washington: Pew Internet & American Life Project.

Luhmann, N., 1980. *Gesellschaftsstruktur Und Semantik: Studien Zur Wissenssoziologie Der Modernen Gesellschaft*. Frankfurt am Main: Suhrkamp.

Mannheim, K., 1952.: *Essays on the Sociology of Knowledge.* (edited by Paul Kecskemeti), London: Routledge and Kegan Paul.

Palfray, J. & Grasser, U., 2008. *Born Digital: Understanding the First Generation of Digital Natives.* Philadelphia: Basic Books.

Ryder, N., 1965. The Cohort ad a Concept in the Study of Social Change. *American Sociological Review*, 30 pp. 843-86.

Andra Siibak

Online Peer Culture and Interpretive Reproduction on the Social Networking Site Profiles of the Tweens[1]

1 Introduction

The new social landscape brought about by the new media technologies has generated a discussion about generational differences in new media use. The visible generational differences are not only connected to the amount of time spent using the new media (Hasebrink et al., 2008) but also in terms of the main opportunities experienced by children and adults online. The main opportunities experienced by children online across Europe are connected to the use of Internet as an educational resource, for entertainment games and fun, for searching global information and for social networking whereas other online opportunities, for example, user-generated content creation or civic participation are much less often practiced (Hasebrink et al., 2008). Compared to the adults who value Internet mainly as an educational resource or an opportunity to gain access to global information; young people priorities the opportunities connected to various forms of online communication, entertainment and play (Hasebrink et al., 2008). Other online opportunities like user-generated content creation and civic participation, however, are less often used.

Mainly because of the early adoption of the new technology and the amount of time spent in front of the computer, the present day children have often been defined by their relationship to technology, as a variety of labels, such as `digital generation` (Papert, 1996), the `Net generation` (Tapscott, 1998), the `digital natives` (Prensky, 2001) or the `electronic generation` (Buckingham, 2002) are connected with new media technologies. These labels are also used to signify the preferences and supposed common characteristics of this generation. Furthermore, as claimed by David Buckingham (2008, p.13), these advocates of the concept of the digital generation regard technology as a liberating force for

---

[1] The preparation of this paper was supported by the research grant No. 6968 and the target financed projects No. 0180017s07 and No. 0180002s7. The author is also thankful for the support of the project "Construction and normalization of gender online among young people in Estonia and Sweden", financed by The Foundation for Baltic and East European Studies.

young people which helps to create a generation that is more open, democratic, creative and innovative than any other generation before them. Nevertheless, theses classifications have not been unanimously accepted by all. For example, Susan Herring (2008) has taken a critical stand about classifications of generational digital divide by suggesting that adults, especially journalist, researchers and new media produces, created the construct of an Internet Generation. She also problematises the severe discrepancy in the adult constructions of this new generation. On the one hand, the mainstream media messages often create moral panics about the possible dangers and risks in the online environments. On the other hand, the majority of new media research as well as advertising campaigns of the new media production companies describe the new Internet generation as novel, powerful and transformative. Therefore for Herring (2008) the various classifications for the new generation reflect the interpretation of a demographic that was not used to growing up with digital media, but not by the today's youth themselves who take digital media for granted. Other theorists (Buckinghan, 2006) have mainly criticised applying too-powerful role to technology and a particular medium as such.

Furthermore, the process of defining a concept like `generation`, however, is a complicated matter as it raises questions about the structure and agency (Buckinghan, 2006). Some researchers (Buckingham, 2002) have been using the concept of generationing in order to differentiate between children and adults. These differentiations are done mainly on the grounds of age, not paying enough attention to the developmental differences of growing up.

According to Mannheim (1952) generations need not be formed strictly in terms of the biological age but it also depends on how people interpret their life chances and how do they see themselves as having a shared identity. Therefore, Buckingham (2006) has claimed that although generations may just be natural phenomena which occur because of the passing of the time, they can also be formed because their members, and possibly non-members as well, have defined meanings of generational membership. Hence, it is possible that some generations may start to demand greater social significant as the members of these generations are more self-reflexive and self-conscious than others (Buckinghan, 2006). Hereby, I prefer to view the new generation of Internet users' through the lenses of self-socialisation (Fromme, 2006). The members of the digital generation take active part in their socialisation process and do not rely entirely on the guidance of adults.

The present article analyzes the use of online social networking sites (SNS), one of the most popular means of communicating online, by 11-12 year olds'. The author views these popular online platforms as new versions of peer culture that are used by the tweens i.e. `the kids on the brink of their teenage years` (Fairchild Teens and Tweens Conference White Paper 2002, cited in Cook &

Kaiser 2004, p. 224) both for the identity constructions as well as for "anticipatory socialisation" (Merton, 1965).

1.1 Peer Culture's Influence on Identity Constructions

Childhood sociologist William Corsaro (1992, p.162) has stated that `childhood socialization must be understood also as a social and collective process` as the first series of peer culture, i.e. `a stable set of activities or routines, artefacts, values and concerns that children produce and share in interaction with peers` (p.162) is produced collectively with the peers. Gaining control of one's life and sharing this control with the others are the two basic themes in peer cultures. Children, however, appropriate the control through play and games that serve a crucial role in their peer cultures.

In the present day context these initial peer cultures are not only formed in play groups and nursery schools, as Corsaro proposed, but also in the online world while communicating and playing together in different online environments. As children do not fully grasp the social knowledge of the adult world, they try to `creatively appropriate information from the adult world to produce their own unique peer cultures` (Corsaro, 1992, p.168). In this way children also become part of the adult culture as they reproduce the elements and information gained from the adult world in their own creative manner.

While communicating in offline as well as online worlds people are always trying to obtain information about each other, in order to be able to know in advance what to expect and what kind of response to give. Erving Goffman was the first to emphasise the importance of impression management i.e. people often engage in activities in order `to convey an impression to others which it is in his interests to convey` (Goffman, 1990, pp. 4). Furthermore, individuals tend to accentuate and suppress certain aspects of the self depending on the context of situation.

Whenever other persons are present, people tend to accentuate these aspects of the self that typically correspond to norms and ideals of the group the person belongs to, or wishes to belong to. Therefore, in order to find out what kind of qualities and features are thought to be sought by potential partners a person may have to `perform` several acts before receiving the approval they were looking for. In case of communicating online the impression management is formulated into a constant worry of how to construct ones virtual identity so that it would be appreciated and accepted among one's peer group.

Jay L. Lemke (2008, p.20) also claims `our identities are the product of a life in a community`, because due to the interaction with various people we are `in the process building up a cumulative repertoire of roles we can play, and with them of identities we can assume` (p.20). Lemke (2008, p.24) proposes that people have two kinds of identity concepts: 'identity-in practice' and 'identity-in-timescales'. The former of which is similar to the Goffman's theory about the identity performances and is used to refer to identity constructions on the short timescales that take place as small-group activities (e.g. playing role-playing computer-games or participating in Internet communities). The latter is a more long-term identity that is not determined by a single identity performance in a single situation, but is made up of several actions and different types of situations we encounter and therefore connected to our *habitus*. Both of these identity notions cannot exist without the other as the two different views upon the same concept are interchangeably linked together. It could be claimed that while constructing profiles for a social networking site, young people practice several identity performances that need not be taken up for a long period of time. Therefore, constructing online profiles could also be viewed as an 'auxiliary ego' (Moreno, 1978) which was created in psychodrama in order to allow the protagonist to see 'oneself' from aside (Moreno, 1978, p.603).

According to Marlen Charlotte Larsen, who used the identity notions proposed by Lemke to analyze the identity performances of youngsters in a Danish social networking site Arto, youngsters are using 'many different identity performances which are all linked to the individual, social, and historical lives of Arto users' (Larsen, 2007, p.16).

Larsen claims that in the context of social networking websites youngsters use their friends and the feedback received from them as 'mediational means' in order to reconstruct one's identity (Larsen, 2007, p.16).

1.2 Social Networking Sites as Spaces for Peer Culture

Social networking websites have become one amongst many online playgrounds available for present day children in the Internet. These online environments are still quite new phenomena however their popularity and influence is constantly growing. Besides the millions of people who play on the virtual playgrounds like

MySpace, Friendster, and FaceBook in order to present themselves, interact with friends or socialise with new people, there are also social networking sites which are meant to connect people with common language or nationality (*Rate* in Estonia, Lunastrom in Sweden, Arto in Denmark etc), common geographical background (Blacksburg Electronic Village), professional background (Linked In), etc. Nevertheless, the basic idea behind all of these websites is universal. Social networking websites not only offer a convergence among the previously separate activities of email, downloading videos or music, diaries, and photo albums; but through these means create an opportunity for self-expression, sociability and creativity for millions of people. Furthermore, according to Sonia

Livingstone (2008, p. 394) 'creating and networking online content is becoming an integral means of managing one's identity, lifestyle and social relations'. Our interest in social networking sites is related to the fact that young people consider these new platforms as "'their' space, visible to the peer group more than to adult surveillance" (Livingstone, 2008, p. 396). Without the recognizable surveillance of adults, children not only start to 'explore the social matrix of relating to others' but they also feel safer when trying out and displaying different constructions and reconstructions of their identity (Ota et al., 1997, p.21).

2 Method and Data

In the following sections I apply the theoretical framework to the analysis of data drawn from an empirical study analyzing the *Rate* profiles of 11-12 year old boys and girls. Children in this age usually step out of relatively peaceful developmental phase (in psychoanalytical tradition called latent phase) and enter a phase of preadolescence, where all the aspects of personal development (e.g. sexual development, social relations with peers, gaining independency from parents etc) are extremely vivid and rapid. My intention was to capture this borderline between 'childish' and 'adolescent' games.

The search engine in *rate.ee* was used for sampling. The age range and gender were inserted in the engine in order to find profiles of youngsters belonging to the 11-12 age group. The engine displayed only the first 300 girls or boys, depending on the search, who were currently online and whose age matched the search criteria. Random selection was used for compiling the sample of 11- to 12-year-olds who happened to be online at that moment.

Search engine displayed 30 profiles on a page from which every first profile was selected for the analysis. My sample consisted of children who were active users of the website and were therefore well aware of the expectations and norms expected to be fulfilled in order to be fully accepted among the community of users. All in all 10 profiles of girls and 10 profiles of boys were selected. Six girls and five boys from the sample were 12 years old, four girls and five boys 11 years old.

Content analysis was used in order to analyze the data provided in the textual parts of the profiles. The main focus of the analysis was to find out how much personal information the tweens displayed on the profiles as well as what kind of hobbies, interest and tastes they proclaim to have. Furthermore, I also analyzed the communities the tweens belonged to.

3 Results

3.1 Socio-Dramatic Role-Play on the Profiles

The analysis of the textual profiles was made in order to analyze the interests the tweens proclaim to have. In case of the girls, 21 categories of interests were created based on the 179 different interests expressed by the girls. In case of the boys, 19 categories included 160 interests enlisted on the profiles of young men. It must be added that all of the labels that stand for the interests are created by the users of Rate. Users can either use the search engine to search for the interests that they would like to include on their profile, or create a totally new label to express their particular interest.

One of the most popular categories among the interests' of the tweens was named 'Rebellious youth'. The labels belonging to the category were made up of different rebellious expressions that could be regarded as part of the socio-dramatic role-play suggested by Corsaro (2003). The tweens relish taking on and expressing power with the use of expressions like 'Tell.it.straight.Up.To.My .Face.Not.Speak.Behind.My.Back.', 'schoolpetrolmatchesbomb'; 'IamwhatIamI fYouDon'tLikeItThenPissOff', or 'Itisbettertoruinmyyouththannottouseitatall', etc[2]. These communities are created in order to develop the 'strategies of resistance'(Cobb et al., 2005, p. 6) against the imposed restrictions, regulations and rules the tweens have to submit to in their life-worlds. While belonging to these communities the young are performing the images of rebellious youngsters who are just trying to go past the rules created by their parents and teachers in the offline society as the online world gives them more freedom to test the boundaries between the right and wrong. The young seem also very much aware of the freedom of expression the online world gives them, as they see these kinds of platforms only as a place for themselves without the restrictions made by the parents or teachers.

Furthermore, the tweens in the sample had also joined in several communities that could be viewed as online spaces of sociodramatic role-play. The majority of communities of this kind e.g. 'Help, my parents won't let me live!', 'We do not break the rules, we just make them ourselves', 'I just like forbidden things', reflected the rebellious lifestyle of the young. By joining these communities, the tweens seem 'use the dramatic license of imaginative play to project to the future - a time when they will be in charge and in control of themselves and others' (Corsaro & Nelson, 2003, p. 112).

---

2   The sayings are translated from Estonian and spelled in a way as it was done by the young.

Corsaro (1997) has also discussed the development of sub-groups and hierarchies inside the sub-groups of children which have formed due to the resistance to rules imposed by adults. The young organise these groups so as to govern themselves and maintain social order in the group. For instance, there were communities in *Rate* where young users could take a role of authoritative figures in order to give advice and share their opinions and pieces of wisdom with others. The agency of the tweens is visible in communities like `Tough man does not smoke`, `All girls are worthy to be treated like princesses`, 'Life is too short to waist it on people who do not care about you`, etc where children are seriously engaged in moral issues to influence their peers. Besides the communities where the young could practice the feeling of empowerment from taking up adult roles there were several communities where the tweens represented different aspects related to being young and the lifestyle of youngsters. Communities like `Music is our freedom`, `MSN is our freedom` or `Thinking left and laughing all the time` were popular among both of the sexes. In comparison to the communities where power and control are exhibited there were also communities which emphasised their young age and problems connected to being young (e.g. `Help, I'm a minor!`, `I hate math!`, `We are the ones who get the strange looks`).

In comparison to the communities and interests that emphasised being in control and empowerment, the communities where some of the girls had joined demonstrated clear evidence of the social ambiguities regarding maturity; sexuality and gender the tweens experience (see Cook & Kaiser, 2004). These communities that were described by sayings like `I am sorry I can't be perfect`, `Sorry that I exist` or `I am ugly` refer to the pressures especially young girls feel in their everyday lives. Joining these kinds of communities can on the one hand be viewed as public declarations of low self-esteem which seem to be vivid result of the expectations of peers and family members as well as the existing beauty norms of the society. On the other hand, the interviews with young experts showed also other possibilities in decoding these communities (see Siibak & Ugur, 2010).

3.2 Construction of the Gendered Self on the Profiles

Studies have referred to the `specialised relationship with friends` (Youniss, 1999, p.23) that the children have inside their peer cultures. Therefore it was no surprise to find that the importance of friends was also stressed under the interests' section of the profiles by both the girls and the boys. In some of the cases the tweens used expressions like `I think that friends are all that matters` or just mentioned being interested in `friends`. The importance of love and close relationships in the lives of the young is also visible in confessing their true feelings of devotion with different sayings like 'if-you'd-only-knew-how-much-you-mean-to-me' or 'When I say I care I really do' or just by including different emoticons that express love (L), or hugging (K) as their interests. Compared to the younger children, the tweens are said to have more stable concepts of friendships and therefore when forging 'social alliances and secure friendship relations they also separate themselves from others' (Corsaro, 1999, p.46). This kind of separation was also visible in case of *Rate* profiles where the tweens in the sample enlisted the names (e.g. Karl, Sandra, Kertu) or nicknames (e.g. Ellu, Raku) of their friends in order to emphasise the special role these friends have in their lives.

Typical heterosexual norm is represented on the profiles of both sexes as in all of the profiles that presented some kind of a romantic interest; it was expressed towards the opposite sex. For instance, boys were emphasizing their interest in 'girls', 'babes' and the opposite gender in general and girls declared being interested in 'boys'. The sex-life and interests connected to the physical love-making were yet not extremely popular, however, already represented as interests in some of the profiles. For example, two of the girls had inserted a saying 'fancyakiss?come&ask&I'llgiveyou' which would hint at a more physical relationship and two of the boys confirmed their interest in kissing and sex.

Gender differentiation in peer cultures is said to reach its peak also in the period of preadolescents (Corsaro, 1999). Differences among genders on the SNS profiles were exhibited for instance when speaking about one's likes and dislikes. Compared to the boys, girls more frequently enlisted the names of different movie stars, singers, rock bands, and other celebrities in their interests section. Boys, however, showed more interest in naming their preferences in terms of music-styles, films, TV-shows, etc.

Gender-specific taste was also displayed while naming the things and objects the tweens liked. For example, girls are interested in more girly-stuff like chocolate, strawberries, photos, the sun, compared to the typically masculine tastes of boys which include things like computers, msn-messenger, Internet, money, bicycles, etc. In comparison to the girls, boys also emphasised stereotypically masculine interests in motor vehicles and sports, the categories which were rarely mentioned by the girls. Furthermore, favourite activities of the girls differed greatly from that of the boys. When girls proclaimed to be interested in very feminine hobbies like drawing, singing, dancing, writing, etc., boys preferred sleeping, jumping, watching TV or listening music.

4 Conclusion and Discussion

In this paper I argued that the identity play of the tweens on the social networking website profiles could be viewed as collective reproductions of children's peer culture. I used the case-study of 11-12 year old users of *rate.ee*, the most popular social networking site in Estonia, in order to analyse how the tweens construct their virtual selves on the profiles. The elements of socio-dramatic role-play and gendered play which were embodied and visualised on the profiles allow me to postulate that social networking websites serve as one of the main opportunities the digital youth use for discovering their aspirational social identity. Furthermore, the elements made up of verbal chanting (e.g. types of self descriptions made upon the play between letters) and other types of collective fun and humour (e.g. belonging to certain communities) that Corsaro (1992) has mentioned as aspects of the peer cultures are clearly visible on the textual parts of the profiles in *Rate*. The analysis of profiles also gives a reason to state that these identity performances of the tweens are partly reproductions of the adult culture that are `generated spontaneously, produced routinely, and shared communally within the peer culture` (Corsaro, 1992, p.162). The tweens seem to carefully monitor the profiles of older users of the websites in order to see what kind of rules and norms should be met while making their own profile entries. In order to get acceptance by the whole community, the tweens consciously make use of the common symbol system recognised by the peers. Thus, it could be stated that the young appropriate their online identity games `to fit with the values and concerns of their peer culture while simultaneously developing social, cognitive, and communicate skills` (Evaldsson & Corsaro, 2005, p.176). The tweens seem to be viewing the medium as a place of their own as they seem more relaxed and open about trying out various identities. The analysis of the textual profiles of 11-12 year olds' give a reason to state that the tweens are ready to explore and test the boundaries between the accepted and the unaccepted among the community of online peers. Although many adults may perceive some of the components of these online identities of the young as not suitable or even outrageous, the tweens themselves just seem to appreciate the opportunity to experiment in order to be able to build up an identity-across-timescales. Furthermore, according to Corsaro and Nelson (2003, p.222) `children's literacy abilities are often enhanced by social interaction with peers`. The influences of peer cultures' style and net speak are clearly visible on the textual parts of the profiles of *Rate* users. It could be postulated that `young children actively contribute to their own literacy acquisition, as well as the literacy acquisition of their peers, by creatively using their skills and abilities to produce written and artistic artefacts that reflect their cultures` (Corsaro & Nelson, 2003, p.223). Hence, social networking websites could also be regarded as one of the favourite places for online content creation for young people. The language play taken up on the textual profiles of the young is full of creativity however the entries may not necessarily be viewed in a positive way in terms of formal education. Rather than supporting the creative self-presentations of the young, I am afraid that the style and form of these online messages would be categorised as perfect examples of deficient literacy skills in the majority of schools (Gilmore, 1986). In order to at least slightly diminish the differences in understandings among the digital generation and others, reverse socialisation (Hoikkala, 2004) is needed.

Bibliography

Buckingham, D. 2008. Introducing Identity. In: D., Buckingham, ed. *Youth, Identity, and Digital Media*. Cambridge, MA: The MIT Press, p. 1–24.

Buckingham, D., 2006. *Is There a Digital Generation? Digital Generations. Children, Young People, and New Media*. Lawrence Erlbaum Associates.

Buckingham, D., 2002. The Electronic Generation? Children and New Media, *Handbook of New Media: Social Shaping and Consequences of ICTs*. Sage.

Cobb, C. L., Danby, S. & Farrell, A., 2005. Governance of children's everyday spaces. *Australian Journal of Early Childhood*, 30(1), pp.14-20.

Cook, D. T. & Kaiser, S. B. 2004. Betwixt and Be Tween. Age ambiguity and the sexualization of the female consuming subjekt. *Journal of Consumer Culture*, 4(2), pp. 203-227.

Corsaro, W. A. & Nelson, E. 2003. Children's Collective Activities and Peer Culture in Early Literacy in American and Italian Preschools. *Sociology of Education*, 76(July), pp. 209-227.

Corsaro, W. A., 2003. *We're Friends Right?: Inside Kids Cultures*. Joseph Henry Press.

Corsaro, W. A., 1999. *Preadolescent Peer Cultures. Making Sense of Social Development*. London: Routledge.

Corsaro, W. A., 1997.*The sociology of childhood*, Pine Forge Press.

Corsaro, W. A. 1992. Interpretive Reproduction in Peer's Cultures. *Social Psychology Quarterly*, 55(2), pp.160-177.

Evaldsson, A. & Corsaro, W. 2005. Play and games in the peer cultures of preschool and preadolescent children: an interpretative approach. In: Jenks, C. (Ed.) *Childhood. Critical Concepts in Sociology*. London: Taylor & Francis, pp. 153-185.

Fromme, J.2006. Socialisation in the Age of New Media. *MedienPädagogik.* Available at: http://www.medienpaed.com/05-1/fromme05-1.pdf [Accessed 12 March 2009]

Gilmore, P., 1986. Sub-Rosa Literacy: Peers, Play, and Ownership in Literacy Acquisition. In: B. B. Scieffelin & P. Gilmore eds. *The Acquisition of Literacy: Ethnographic Perspectives.* Norwood, NJ: Ablex, 155-168.

Goffman, E., 1990. *The Presentation of Self in Everyday Life.* Penguin Books.

Hasebrink, U., Livingstone, S. & Haddon, L. 2008. *Comparing Children's Online Opportunities and Risks across Europe: Cross-National Comparisons for EU Kids Online.* EU Kids Online (Deliverable D3.2).

Herring, S. C. (2008). Questioning the generational divide: Technological exoticism and adult construction of online youth identity. In: D. Buckingham (Ed.), *Youth, identity, and digital media* Cambridge, MA: MIT Press, pp. 71-94. Available at: http://ella.slis.indiana.edu/~herring/macarthur.pdf [Accessed 3 February 2009]

Hoikkala, T., 2004. *Global Youth Media and New Forms of Socialization. In: United Nations Workshop on Global Youth Culture.* New York.

Larsen, M., C., 2007. *Understanding Social Networking: On Young People's Construction and Co-construction of Identity Online.* In: *Internet Research 8.0: Let's Play,* Association of Internet Researchers, Vancouver.

Lemke, J., 2008. Identity Trouble. *In:* C. R. Caldas-Coulthard & R. Iedema, R. eds. *Critical Discourse and Contested Identities.* Basingstoke, England: Palgrave Mcmillan.

Livingstone, S., 2008. Taking Risky Opportunities in Youthful Content Creation: Teenagers' Use of Social Networking Sites for Intimacy, Privacy and Self-Expression. *New Media & Society*, 10(3), pp. 393- 411.

Mannheim, K., 1952. The Problem of Generations. In: Mannheim, Karl (edited by Paul Kecskemeti): *Essays on the Sociology of Knowledge.* London: Routledge and Kegan Paul pp. 276-320.

Merton, R. K., 1965. *Social Theory and Social Structure.* New York: Free Press.

Moreno, J. L., 1978. *Who shall Survive?* Reprint of 2nd ed. Beacon, NY: Beacon House Inc.

Ota, C., Erricker, C. & Erricker, J., 1997. The Secrets of the Play Ground. *Pastoral Care in Education,* 15(4), pp. 19-24.

Papert, S., 1996. *The Connected Family: Bridging the Digital Generation Gap.* Atlanta, GA: Longstreet Press.

Prensky, M., 2001. Digital Natives, Digital Immigrants. *On the Horizon*, 2001, 9(5), pp. 1-2.

Tapscott, D., 1998, *Growing up Digital: The Rise of the Net Generation*. New York: McGraw-Hill.

Youniss, J., 1999. Children's Friendships and Peer Culture. In M. Woodenhead, D.

Faulkner, & K. Littleton Eds., *Making Sense of Social Development*. London: Routledge, pp. 13-26.

Mutlu Binark and *Günseli Bayraktutan Sütcü*

Usage Patterns of New Media by Turkish New Middle Class Young People

1 Introduction: Features of the New Middle Class Youth Culture in Turkey

In this paper, various usage patterns of new media in everyday life by the new middle class Turkish youth are conceptualised within the context of a consumption culture. The usage of new media within the context of consumption culture, basically means that individuals utilise information technologies within cultural and social contexts (Slack & Macgregor Wise, 2002; Hine, 2000). Data from the United Nations' Information Economy Report 2006 shows that the use of mobile phones was more widely spread than Internet access in households in Turkey in 2005: while 16 million people had access to the Internet, either from public places or from home, the number of people who had at least one mobile phone was over 43,5 million in the same year (2006, pp.56-70). Recent research findings on the usage of household information technologies, conducted by the Turkish Statistical Institute in April 2009, show that 30% of households in Turkey have Internet access; 85.6% of this access via Broadband. According to this research, it is predominantly those in the 16-24 age group that use Internet and computers. From January to March 2009, this group used the Internet for the following purposes: 72.4% of them for sending-receiving e-mails; 70% of them for reading online e-newspapers; 57.8% of them for chat and instant messaging; and 56.3% of them for playing digital games and downloading movies. (Radikal, 2009, p.4). Based on this data, it could be suggested that young people mostly use new media facilities for communication to support their interpersonal relationships. Before exploring these new media usage patterns, the diverse lifestyles of young Turkish people need to be reviewed.

Youth identity and its culture in Turkey are constructed within a framework of various ideologies. One such ideology is Kemalism, the essence of which is a combination of secularism, modernism and liberalism, with a strong emphasis on the unity of the nation-state. Another prominent ideology, conservatism, combines anti-EU and anti-globalist nationalist ideology. A third ideology concerns the identity politics of Islamist religious discourse. The secularist nationalist leftist ideology is also significant. Lastly, secular socialist ideology should be included. There are sub-divisions of these main ideologies, which are stimulated by other social, political, and economic factors, such as national political conjunctures and regional political conditions (Kentel, 2005; Lüküslü, 2009; Yılmaz, 2007). Despite the existence of different ideologies in the Turkish political environment, it can be said that young people in Turkey have been mostly mobilised by two contradictory ideologies: hegemonic modernist liberal ideology and conservative values and beliefs. Youth cultures are also determined by two different social and economic situations: one is urban middle and upper class consumption practices, and the other is the everyday life practices of the lower and underclass. Social and economic inequalities fragment young people into different and unrelated social networks (Kentel, 2005, p.14). The middle and upper class youth – the generation born after 1980, who are also the subjects and interviewees for this paper – have grown up within a neo-liberal ideology and its cultural values. Raised in a privileged environment, they are able to draw on their parents' economic capital in order to acquire symbolic capital and status. They are accustomed to a life within a material culture characterised by brand names and access to this culture becomes a marker of individual taste. Their life is thus centred on the enjoyment of the latest products of the culture industry. Recent consumption trends among these young people include alternative sports, healthy foods, alternative healing techniques, the development of interpersonal skills, digital games, and spending time in malls, as well as spending time and money on appearance, as dictated by the global culture industry. In line with Pierre Bourdieu (1977), it could be suggested that the consumption practices of the new middle class in Turkey not only reflect social borders and differences, but also construct symbolic borders between the lower and middle classes. As a result, the new middle-class persona and his/her ethics are crystallised into a desire to consume, a growing tendency towards individualism, ironical indifference, hedonism and narcissism. They gain knowledge and experience through formal education, which constitutes their cultural capital, and then they mostly take up careers in the following fields: banking, media, public relations and telecommunications. As Meltem Ahıska and Zafer Yenal argue (2006, p.62), the new middle class consists of the privileged children of neoliberalism in Turkey.

2 Lifestyle, Consumerism and the New Middle Class: a Brief Review of the Literature

Today people are categorised mostly by their symbolic and material consumption patterns (Pilkington & Johnson, 2003, p.265), so lifestyles and consumption patterns are important factors when discussing young people's usage of new media. Lifestyle is both an expectation and a decision about how one locates oneself within certain social, cultural, and economic structures. Today, lifestyle is closely related to individualism, consumerism, and the performance of these two practices in front of other people (Pilkington & Johnson, 2003, p.265). In his discussion of the concept of lifestyle, Mike Featherstone makes the following comment:

> While the term has a more restricted sociological meaning in reference to the distinctive style of life of specific groups (Weber, 1968; Sobel, 1982; Rojek, 1985), within contemporary consumer culture it connotes individuality, self expression, and a stylistic self-consciousness (Featherstone, 1991, p. 83).

Featherstone claims:

> consumption, then, must not be understood as the consumption of use-values, a material utility, but primarily as the consumption of signs". So, lifestyle is also related to the formation of new middle-class personas(Featherstone, 1991, p. 85).

According to Stephen Edgell(1993, p. 62), the new middle class – also known as the small or petite hayırbourgeoisie and the white-collar or non-manual classes – is property-less. But today this definition alone does not fully explain the complexity of their everyday life practices, especially when the usage of new media is taken into consideration. Bourdieu argues that the new petit bourgeoisie has certain features such as setting the standards for good taste, being new culture intermediaries, while occupationally located in service sectors like finance and marketing. This conceptualization of the new middle class is imbricated with the theoretical discussions on consumption and lifestyle.

As individualism is one of the key determinants of the new middle class persona, how this persona locates him/herself within society ought to be explained. Frances Bonner and Paul du Gay discuss (1992, p.86) the individualism of new a middle class persona as follows:

> Under this new 'regime of the self' consumers are constituted as individuals seeking to maximize the worth of their existence by assembling a lifestyle, or lifestyles, through personal acts of choice in the marketplace. So the 'right' to choose entails a parallel 'obligation' to make the most of your own individual existence.

This individual new persona has at least three basic characteristics: an imagined self rather than a true self; superficial personality, beliefs and activities; and fragmentation of coherence (Löfgren, 1994, p.49). These three characteristics also define the new middle class youth culture in Turkey. These new personas have limited cultural capital: their expertise derives from their ability to use this limited cultural capital efficiently. So, they can be called 'good players', or better still, 'bluffers'. Featherstone clarifies (1991, p.91) this performance as an approach to life which is characterized by a 'why can't I have my cake and eat it?' attitude quest for both security and adventure.

Similarly, members of the new middle class in Turkey demand both adventure and security at the same time. Through the opportunities offered by the new media, young people become *net flaneurs,* embarking on adventures without taking any real life risks.

The appearance of such lifestyles is related to the shift in consumption patterns from the satisfaction of basic needs to the search for quality and taste. Consumer culture drives the new persona to invest in the knowledge of how to distinguish between new and old trends. The new persona is constructed upon permanently unsatisfied desire and dwells in false consciousness concerning the satisfaction of this desire through consumption. The emergent persona of this new middle class always asks the question 'Who am I?' instead of 'How should I live?' As Manuel Castells puts it, "People increasingly organise their meaning not around what they do, but on the basis of what they are, or believe they are" (1996, p. 3). Castells relates the usage of new information technology to the individual's identity construction. The new middle class personae are associated with an urban identity, therefore they look for people with similar origins. As a consequence, the new middle class live in closed groups, isolated from other classes. Vis-à-vis this theoretical conceptualization, Serhat Gürcü[1], the Manager of *Youth Republic Research and Advertising Company,* an İstanbul-based research company, analyses young people's consumption trends and describes the basic characteristics of the Turkish youth as follows:

> In Turkey, the young people's average yearly pocket money is around 2,200 US dollar. In Ivy league private universities, this pocket money rises around 7000 US dollar. Such an amount is more than gross national income. It doesn't matter how much money these youngsters have. They prefer to have the most expensive mobile phones, because of prestige. For them, mobile phone shows what kind of person you are, when you put it on table. Young people easily persuade their parents to buy technological items. I could explain the characteristics of these youngsters with this example: They define themselves with brand names, such as "I am like Mango", "I am more Adidas", "I am Puma style person" etc. They have consumption-centreed life styles and they have high expectations from the future, but on the other hand they don't like to work for this. Somehow, they think they deserve to have a lot of money and high social status. They have individualized life styles, but they could quickly organize, due to network effect, such as using Internet. They have 24 hours MSN on. I could definitely say that they are not idealist...Their common anxiety is career.

---

[1] The interview was held in August 22nd, 2007, in İstanbul. The web address of the company is hhtp://www.youthrep.com.

3 Research Methods and Research Questions

Empirical and quantitative research is limited[2] regarding the usage patterns of new media by young people, based on class, gender, and lifestyles in Turkey. As our study aims to analyse the usage patterns of new media (Internet, mobile phones, digital tv, etc.) by this nascent group of people, we conducted focus group research in the Faculty of Communication at Başkent University[3], a private university founded in Ankara in 1992. University education[4] in a private institution can be considered as one of these isolation strategies because these institutions consist of a homogeneous population in regard to the students' social, economical and cultural backgrounds. Therefore, it is rare to encounter members of the lower classes in private universities, unless they have scholarships.

---

2   Some of these limited researches are following: *CivicWeb Project* (2007-2009), which aims to question the potential contribution of the Internet in promoting civic engagement and participation among young people (aged between 15 and 25). The project is supported by the 6th framework programme of the European Commission. Other studies include: Günseli Bayraktutan-Sütcü and Mutlu Binark (2006). "Screenwide Socialization: So New, So Familiar: Usage Patterns of New Media", a paper presented by European Sociology Association (ESA)-Annual Conference in Antalya, 03 November 2006; and Mutlu Binark and Günseli Bayraktutan Sütcü (2006). "Teknogünlüklerdeki Çoklu Sessiz Yaşamlar: Yeni Medyanın Sessiz Enstrümanları-Yeni Sınıf Gençlik", a paper presented in Yeni İletişim Ortamları Etkileşim Uluslararası Konferansı, 1-3 Kasım 2006, Marmara Üniversitesi İletişim Fakültesi.
3   The university currently has 11 faculties, 7 institutes, 11 research centres, a conservatory, three vocational schools and 93 associate, bachelors, and post graduate programmes. Başkent University also has many subsidiary units in health, media, tourism, textiles and chemicals. The yearly tuition per student is approximately 7000 Euros.
4   According to the results of the latest national census organized by the Turkish Statistical Institute in 2000 (Türkiye İstatistik Kurumu), Turkey's population is 68 million and the number of people aged between 0 and 14 is 20.220 million, people aged between 14 and 65 is 43.701 million and people aged over 65 is nearly 4 million. Thus, it can be said that two thirds of the population is composed of young people. The Turkish Ministry of Education declares that more than 2 and half million students are enrolled in 141 universities in the country. Almost 500 thousand of these university students are registered with the Open University (See: www.yok.gov.tr). Some of these university students are members of the new middle class Turkish youth, which will be discussed within the context of consumption culture in this paper.

A focus group interview is:

> specific type of group discussion which is usually conducted in a FTF (face to face) format involving between five and ten participants (Mann & Stewart, 2002, p.99).

In this field research a focus group interview was carried out with five youngsters selected among the freshman students from the Faculty of Communication in 24 July 2009[5]. In this focus group, one participant was male, the others were female, and they came from different departments of the Faculty. As noted above, the aim was to learn about young people's usage of new media (mobile phones, Internet etc.) and their attitudes towards social networking sites and related issues like privacy and democratic participation. Informant A and E live with their families, C lives in two houses (family house and his apartment), B lives in the dormitory and D lives alone because her family lives in a different city. All of them have mobile phones. Their parents pay their monthly GSM expenditures. They have broadband connection and laptops. On the basis of these findings compared with the theoretical framework discussed above, the interviewees could be defined as future members of the new middle class.

4 Research Findings

For the new middle class youth, school and home are at the core centre of their everyday routines. According to Ferhat Kentel, these two conservative institutions are where young people are inculcated with the dominant values of Turkish society (2005, p.15). In such limited social networking, these young people do not experience different social encounters. They distinguish between school as a public place and home as a private sphere. When we asked them about their daily routine, they mostly answered it on the basis of home-school-home activities. Two important findings regarding the new media usage are: firstly, based on our focus group findings, it can be said that our informants are *homo oeconomicus*. They calculate and make a rational choice among different GSM services. Our informants say:

> "I have one line. Once I transfered my line to Telsim, then returned to Avea. Then changed the line Telsim again. Why I acted like that? I regretted to change from Telsim to Avea. Because of advertisements... When I changed the operator, I understood there was nothing good in Avea. Nowadays, I am considering to change the operator to Turkcell." (A)

---

5 We assigned letters to our interviewees as identification in order to protect their privacy: A (female, 21 years old), B (female, 22 years old), C (male, 24 years old), D (female, 28 years old) and E (female, 20 years old).

"...I have three lines. As you know, based on the different pricing policies...I'll try to pay the mobile phone fee myself."(C)

" I have two lines. One is for the teachers. I use that line for my friends". (D)[6]

Secondly, new media is an indispensable part of their daily life, they transfer their offline relations to the online environment.

"I never turn off my MSN. In deed, computer and the net are parts of my life. How should I tell? I wake up with keyboards, I sleep on keyboards. I use MSN to talk with my friends, I play online games. Without computer and the net, I can not imagine a life. On MSN, I rarely talk with strangers. I usually communicate with my friends, share homeworks etc." (B)

"I use my Facebook account, as a telesecretary. I tell my friends the suitable time to make a call to me etc. As my personal policy, I don't communicate with strangers through the net, except e-trade. Once I had a strange invitation. I mean, the gays invite you to be their friends. Because of these invitations, I closed my Facebook account to outside. I only communicate with my friends through the net... (C)

Serhat Gürcü's observations regarding young people's usage patterns of the Internet, especially MSN contain similarities:

"Their Internet is always on. 95 % of them have MSN account, so they could communicate easily. The main characteristic of these young people is their existence within the tribes like organizations, in other words, the network structure. For instance, an ordinary young people has 66 people in his/her MSN list. This means that if you disseminate only one message, and if you persuade him/her about your sincerity, you could reach 66 people in advance. We call this technological influence as a network affect."

It is possible to categorise the usage strategies of new media in terms of maintaining social networks. Firstly, young people communicate with existing school friends in order to supplement face-to-face communication with them. Secondly, they communicate via new media in order to maintain old relationships. Their socializing can be called *immobile socializing* within known and safe social networks (Bakardjieva, 2003). They always communicate within a social network composed of their previous relations, such as family members, relatives, friends (old and new but established in the context of their real life), girlfriends or boyfriends. They do not contact new people over the Internet. There may be certain reasons for their preference for communicating within known social networks; one possible assumption is that they do not have the so-

---

6 She mentions the advertising campaign of one of the GSM companies. This particular company runs certain advertising campaigns for professionals in the public sphere such as teachers, doctors and army officials, so they can use this GSM line and save money.

cial and cultural resources to make contact with new people. In relation to this issue one of our informants says:

> "I opened a blog once. But, then I didn't find anything to write down. My life is so much mono-coloured. So, I left the blog."(C)

In other words, they do not have anything in their social and cultural repertoire to equip them for new relationships. By restricting their communication they avoid two potential risks: one is to reveal their true identity and the second concerns the insecurity of heterogeneous social encounters. Our research shows that new middle class young people feel the need to control space in order to feel secure. John Field discusses (2006, p.146) this issue in his work on social capital, arguing that people seek out the same social networks consisting of people who have the same ideas and interests even in cyberspace. Therefore they are hostile to strangers who suddenly participate in their forums, chat rooms or MSN. Alesia Montgomery's research into social networking in cyberspace also shows that people prefer to communicate with those they already know from face to face relationships (Castells, 2005, p.481). Facebook is one of the most popular and most visited social networking sites among the interviewees for social bonding and social grooming[7].

The content of the messages sent by mobile phones or via the Internet is connected with young people's daily routines. They have a growing dependency on new media to arrange their daily activities and to maintain their online and offline social communities (McMillan & Morrison, 2006). For example, the interviewees for this research communicate online before and after face-to-face communication. The time they spend on mobile phones is generally related to the arrangement of offline relations. Sms (short message service) and Facebook walls are also used to engage with the social networks established in their real life. The result is that face-to-face relationships are strengthened, because these young people use new media in order to make daily arrangements, to feel secure, and to be relieved of the boredom of their real lives. New patterns of communication also emerge through the use of new media. For example, the interviewees make their friends' mobile phones ring simply to imply 'You're in my mind' or 'Call me', and thus 'real' social networking is once again empowered. In the Facebook environment this pattern is called "poking".

The outcomes of these usage patterns are also observed in their academic research activities. The interviewees are able to use the new media, and especially the Internet for academic purposes, but their searches are restricted to Google; the names of specific databases or dictionaries located on the Internet are never mentioned. Since they construct themselves within such a limited arena, in this

---

[7] Facebook is the third most widely used website by Turkish Internet users with 12.800 million visitors in July, 2009. Available at: http://www.alexa.com

case Google, they also maintain their habits of secure networking even when conducting their (re)search. They do not use the available opportunities to access different and credible academic resources, so they do not develop or mobilise their intellectual capital. Considering the interviewees' need for security and the fact that they prefer to continue their social networks online, it could be argued that these new middle class personas reveal considerable paranoia as well as phobias such as xenophobia[8], which could easily be motivated by nationalist and racist ideology, aiming to eliminate diverse political, social and cultural identities from the public sphere in Turkey.

Young people do not use just one form of new media; rather, they use them in combinations to construct a communication landscape for themselves (Johnsson et.al, 1998). Generally, the latest technology products, the new media our interviewees own/access allow for many other usage patterns beyond their primary function. The interviewees apply the functions of new media in order to follow entertainment and celebrity culture. They search for infotainment about celebrities on websites and mainly visit gossip sites. Graeme Turner claims that consuming celebrity gossip "…offers a way of imaginatively enriching one's community…" (2004, p. 115). Given that the interviewees for this research have limited social networks and that they prefer to transfer their offline social relationships online, Turner's argument also becomes pertinent to this research. In following celebrity gossip through the new media, these young people are acquiring a stock of cultural capital to be shared in their social circles. In addition, they use celebrity culture and entertainment products as wallpapers, ring tones, start-up screens, and screensavers. This inter-textuality indicates the degree of the hegemony of mainstream popular youth culture in Turkey.

> "I follow my favorite pop singers fan groups at Facebook, to learn their concert programmes." (E)

---

[8] Here, we should mention that Facebook is becoming one of the most popular places to disseminate hate speech among Turkish young people. A recent study on the types of social organizations in Facebook opened by Turkish young people shows that there are thousands of Facebook groups whose discourses are xenophobic, homophobic and racist. Most of these groups use the Turkish flag and other nationalist symbols to indicate their pure nationalist stance, vocalize national sentiments, and organise hate speech campaigns against other ethnicities, homosexuals, and foreigners living in Turkey (Toprak, et.al. 2009). For example; the groups' names, "1,000,000 Turk", "1,000,000 fans of Atatürk ", "I will find 1,000,000 Turks against Kurdish Resistance Army" have a racist, xenophobic and militarist discourse; and these Facebook groups produce stigmas against ethnic groups living in Turkey such as Kurdish people, Armenians, Jewish people, and homosexuals. Our informants are aware of these groups but they do not participate in such groups. However, neither are they concerned with hate speech and they do not complain about these activities to Facebook Turkish admins. This attitude demonstrates an apathetic position regarding political and social issues, unless something directly affects their individual interest in social network sites.

As a result, mainstream symbolic signification practices emerge among the new middle class youth in Turkey. These young people locate their habitus within the global consumption culture, which includes some local moral values and traditional cultural patterns. For example they follow international stars and they are fans of music groups:

> "I participated in a group, which organized a tour of Metallica to Turkey. That was a successful campaign." (C)

The second research question concerns young Turks' perception of social networking sites, such as, the most popular one, Facebook. We asked them what they thought of these kinds of web sites and services. All of the people interviewed have a Facebook account. They opened an account to strengthen their offline social relations and maintain these social relations, but they much prefer other instant messaging services. In response to our questions regarding the issue of privacy, they mostly problematised the graphic/visual representation (profile pictures and photography albums) regarding Facebook. One of the interviewees said:

> "Telling the truth, there are many disadvantages at Facebook. One is the increasing amount of advertisements at the environment. The second thing is the excessive amount of video sharing. I really hate. Once you open the main page, you will see hundreds of video share. Now, I heard that the admins are going to put the ads on the beginning of the videos. So we have to watch the ads. Previously, we used to ring the phones, now uploading videos. Third disadvantage is the risk of stealing profile pictures. It is a bullshit. Fourth disadvantage is the labeling. Once you put a picture and labeled it. One could remove these labels, if he/she does not like the picture. Once I had this experience. People want to be seen as if they are Brad Pitt or Angelina Jolie. You are what you are. So, they don't have a right to remove my labels. Once you agreed on to be taken your picture, you agree on labeling at Facebook. People put their best pictures at Facebook, because of ego. I never put my best pictures at Facebook, it is a false consciousness, I mean." (C)

In this research we also question people's opinions regarding the potential of social networking sites as a tool of democracy. Interviewees think that Internet and social networking sites could be used as a means of democratic participation. There are two major reasons for this response, first of all their conceptualization of being political is somehow different. When they were asked; "how you do you define being political/ behaving politically?" one of the people interviewed replied:

> "Being and acting political. I mean, the person who knows how to act and how to behave among people. This person does not cause any misunderstandings, so he or she is political. I am in this sense very political person. I don't get angry to any one; I control everything." (C)

Secondly, they believe that because the offline environment is not democratic, any online democratic process could not survive and affect the offline world. In other words they think that offline democracy is the prerequisite for online democracy.

> "I think the University is administrated by gestapos. So there is no meaning to organize any petition campaign even real life. Don't you remember the spring festival? We organized a campaign…But there was nothing after that. Indeed, this problem is related with the soil of human. Turkish people like to talk, but never do anything. If we had written a petition individually, we could be successful. Via Internet we can't organize and obtain something concretely."(C)

Also in the *Civicweb* project similar findings are discussed:

> "…We have found out that forums are generally available, but are not frequented by youth. Most often, young people prefer mail groups which are more practical to use, require less time and which do not ask for any expertise or specific information for self-actualization unlike forums. Furthermore, youth activists seem to build their networks online, but prefer offline interaction for local and grassroots action."(Telli-Aydemir, 2008, p. 65).

In order to learn about interviewees' opinions regarding democracy we asked them about Internet bans, which is a very topical issue among Turkish people. For example Youtube has been banned by court decision for almost a year, however Internet users could access Youtube content by using different IP's and by developing alternative technical solutions. One of the people interviewed argued that:

> "Coming to Internet bans in Turkey; amateur society's amateur bans. With oppression you cannot reach anything. You cannot forbid thinking or thoughts. For example, thousands of web sites are banned, but one can reach via other ways. So, Internet can not stay any bans…" (C)

People were also asked whether they have participated in any of the online voting processes. One person replied;

> "I used Internet to support TEMA, Altı Nokta Körler Derneği, and Turkey's Eurovision Song Contest by my votes. I also supported my uncle's candidacy at his party's pre-election, conducted through Internet online voting with my votes. I send at least 1000 votes." (C)[9]

Results show that the people interviewed participate in political life but they encounter it as something like business interaction. They send votes and someone becomes the winner in a pre-election. They do not think about the

---

9   TEMA is a foundation for the protection of the environment and Altı Nokta Körler Derneği is an association for blind people.

whole procedure and neither do they have any motivation to participate in the rest of the process. There is no kind of belief or trust in the democratic political procedure and they do not feel any commitment to the democratic political sphere. Demet Lüküslü, also underlines (2009, p.148) that young people in Turkey see politics as a field of corruption and clientelism.

5 Conclusion: The New Middle Class Persona's Invisible Performance in the Public Sphere

Because of new middle class ethics and values in Turkey, the transformation of passive individuals into efficient citizens is both a social and cultural problem. Two important findings from this research could be interpreted as evidence for this argument. First, none of the people interviewed participates in any civic or political activities via new media. The second finding is that they are not present as 'web citizens' on the net. Although they have all learned how to create websites as part of their education in the visual communication design department, none of them has a personal website. Furthermore, they are more trained in the uses of the new media technologies than other students within the university. Despite their cultural capital, none of the people interviewed locate themselves as actors in the new media environment, either socially or politically. As Peter Dahlgren says:

> "This results in a reduced 'social capital' among citizens- seen specifically as diminished networks of social contacts- which includes not least a reduction in communicative competencies.... With increased fragmentation and atomization follows a decline in social trust, which further inhibits participation." (Dahlgren, 2006, p.272).

The people interviewed during this research also have diminished networks of social contacts, which causes a reduction in their communicative competencies and, naturally, in their political participation. This reduction in communicative competencies reinforces their profound silence in terms of their status as social and political citizens. These young people have knowledge of new media literacy, but they do not advance beyond this. The reasons for which these young people use this media are shaped by the new middle class youth culture and its ethics. Although they are not effective political citizens, they can be defined as perfect specimens of *homo oeconomicus* because they frequently calculate the financial consequences of using different GSM services. As Zygmunt Bauman also argues that:

> "...*homo oeconomicus* the lonely, self-concerned and self-centered economic actor pursuing the best deal and guided by 'rational choice', careful not to fall prey to any emotions that defy translation into monetary gains, and populating a lifeworld full of other

characters who share all those virtues but nothing else besides. The sole character the practioners of the market are able and willing to recognize and accommodate is *homo consumens*- the lonely, self-concerned and self-centered shopper who has adopted the search for the best bargain as a cure for loneliness and knows of no other therapy; a character for whom the swarm of shopping-mall customers is the sole community known and needed; a character whose lifeworld is populated with other characters who share all those virtues but nothing else besides"(Bauman, 2008, p.69).

Within the neo-liberal world order, the main global aim is to internalise the logic of the market efficiently, as the interviewees indeed do. Therefore, in the context of consumer culture, young people learn how to assume the persona of the perfect individual, materialist, conformist, who is indifferent and hedonistic and only engages with the same small circle of relations, both online and offline. The interviewees' public silence and indifference to the world and to Turkey's social, political, economic affairs imply that as future members of the new middle class, they are brought up in accordance with neo-liberal ethics and values as the very basis of their identity formation process which, in turn, is in harmony with mainstream signification practices. Neoliberal ethics support the life motto that: "money brings happiness" (Kazgan, 2002, p.170). Because of this political and cultural context/mentality young people internalise these ethical values and behave accordingly and as Bauman mentioned above, the interviewees turn to *homo consumens*. Graham Murdock also explains how the ideology of consumerism affects people:

"...the ideology of consumerism encourages people to seek private solutions to public problems by purchasing a commodity....It also redefines the nature of citizenship itself so that it "becomes less a collective, political activity than an individual, ecomomic activity- the right to pursue one's interests, without hindrance, in the marketplace "(Dietz 1987, p. 5) (Murdock, 1992, p. 19).

The interviewees' performance as citizens ironically occurs through their public silence, while their usage patterns of new media reconstructs them as silent instruments in the public sphere. These new middle class young people use new media to be good members of the global consumption culture. Through new media, they maintain their secure yet isolated social relations far from the lower classes, in other words 'the others,' in Turkey. In addition, they feel insecure due to the rise of social and economical inequalities between the social classes, and the growth of an sub-class population implies a potential threat to their everyday life practices. Therefore, the following questions take on importance: how can we use new media to alter these young peoples' culture, so that they could have the opportunity to encounter different and alternative ways of life? How could new media be used to cultivate a democratic mindset and free thought among mainstream young people? Potential means of altering the mainstream lifestyles of these young people are critical pedagogy, new media literacy, critical solidarity, and responsibility ethics, which should be practiced in formal educational institutions as well as in every other aspect of life. Young people would thereby expand their social and cultural capital, which would ultimately result in the transformation of passive citizenship into participatory citizenship.

Bibliography

Ahıska, M. & Yenal, Z., 2006. *The Person You Have Called Cannot Be Reached at the Moment: Representations of Lifestyles in Turkey 1980-2005*. İstanbul: Ottoman Bank Archives and Research Centre.

Bakardjieva, M., 2003. Virtual Togetherness: an Everyday-life Perspective. *Media, Culture & Society*, 25, pp.291-313.

Baumann, Z., 2008. *Liquid Love: On the Fragility of Human Bonds*. Cambridge: Polity Press.

Bonner, F. & du Gay, P., 1992. Representing the Enterprising Self: thirstysomething and Contemporary Consumer Culture. *Theory, Culture & Society*, 9, pp. 67-92.

Bourdieu, P., 1984. *Distinction: A Critique of the Judgment of Taste*. London: Routledge and Kegan Paul.

Bourdieu, P., 1977. *Outline of a Theory of Practice*. Translated from French by R. Nice. Cambridge, New York: Cambridge University Press.

Castells, M., 1996. *The Rise of the Network Society*. Cambridge: Blackwell.

Castells, M., 2005. *Enformasyon Çağı: Ekonomi, Toplum ve Kültür-Ağ Toplumunun Yükselişi. Birinci Cilt*. Translated from English by E. Kılıç. İstanbul: İstanbul Bilgi Üniversitesi Yayınları.

Dahlgren, P., 2006. Doing Citizenship- The Cultural Origins of Civic Agency in the Public Sphere. *European Journal of Cultural Studies*, 9 (3), pp. 267-286.

Edgell, S., 1993. *Class*. London: Routledge.

Featherstone, M., 1991. *Consumer Culture and Postmodernism*. London: Sage Publications.

Field, J.,2006. *Sosyal Sermaye*. Translated by B. Bilgen & B. Şen. İstanbul: İstanbul Bilgi Üniversitesi Yayınları.

Hine, C., 2000. *Virtual Ethnography*. London: Sage.

Johnsson-Smarapdi, U., D'Haenens, L., Kratz, F., & Hasebrink, U., 1998. Patterns of Old and New Media Use Among Young People in Flanders, Germany and Sweden. *European Journal of Communication*, 13(4), pp. 479-501.

Kazgan, G. ed., 2002. *Kuştepe Gençlik Araştırması*. İstanbul: İstanbul Bilgi Üniversitesi Yayınları.

Kentel, F., 2005. Türkiye'de Genç Olmak: Konformizm ya da Siyasetin Yeniden İnşası. *Birikim*, 196, pp.11-17.

Lasch, C., 1978. *The Culture of Narcissism-American Life in An Age of Diminishing Expectations*. New York: W.W. Norton&Company.

Löfgren, O., 1994. Consuming Interests. In: J. Friedman ed. 1994 *Consumption and Identity*. Switzerland: Harwood Academic Publishers.

Lüküslü, D., 2009. *Türkiye'de 'Gençlik Miti': 1980 Sonrası Türkiye Gençliği*. İstanbul: İletişim.

Mann, C. & Stewart, F., 2002. *Internet Communication and Qualitative Research- A Handbook for Researching Online*. London: Sage Publications.

McMillan, S. J & Morrison, M., 2006. Coming of Age with the Internet. *New Media & Society*, 8(1), pp.73-95.

Murdock, G., 1992. Citizens, Consumers, and Public culture. In: M. Skovmand, & K.C.Schroder ed. 1992 *Media Cultures: Reappraising Transnational Media*. London: Routledge, pp.17-41.

Pilkington, H. & Johnson, R., 2003. Peripherial Youth: Relations of Identity and Power in Global/Local Context. *European Journal of Cultural Studies*, 6(3), pp. 259-283.

Radikal, 2009. 19.08.2009, p.4.

Slack, J.D. & Wise, J. M., 2002. Cultural Studies and Technology. In: Leah A. Lievrouw & Livingstone, S., ed. 2002 *Handbook of New Media*. London: Sage. pp. 485-501.

Telli-Aydemir, A., 2008. Civicweb: Youg People, The Internet and Civic Participation, Citizenship and Cultural Identities in the EU: Old Questions, New Answers. International Conference in İstanbul, 18-19 October 2007, ESSHRA, Ankara, TÜBİTAK, pp. 57-70.

Toprak, A. et al., 2009. *Toplumsal Paylaşım Ağı Olarak Facebook: "Görülüyorum Öyleyse Varım!"*. İstanbul: Kalkedon.

TÜİK, 2009. *Kültürel Faaliyetlere Katılım ve Ayrılan Zaman 2006*. Ankara: Türkiye İstatistik Kurumu Matbaası.

Turner, G., 2004. *Understanding Celebrity*. London: Sage Publications.

United Nations., 2006. *Information Economy Report*. Geneva: UN Publications.

Yılmaz, H., 2007. *Türkiye'de Orta Sınıfı Tanımlamak*. İstanbul: Boğaziçi Üniversitesi-Açık Toplum Enstitüsü.

Ariela Mortara

Generations and Media Fruition of Social Networks

1 Introduction

The participative logic of Web 2.0 and the spreading phenomenon of social networks are increasingly attracting the attention of academics and researchers. This is the case even though internet social networks share some of the features of all social constructs and consist of a group of people sharing some sort of bond, whether that may be a family bond or something less strong. There are many studies on relationships and bonding within the same social group, some going back to the 1800s. One important author, Durkheim, (1971) reflected upon the difference between a traditional society in which mechanical solidarity allowed interaction among people sharing a collective conscience and an advanced capitalistic society. The latter, the modern industrial society, requires moral rules to control the cooperation of very different individuals and only organic solidarity can confer roles and tasks to them.

Tönnies (1979) compared personal and social relationships characterizing Gemeinshaft to impersonal relationships, typical of Gesellschaft, arising in urban and capitalist settings.

Through time, social network analysis (Scott, 1988) has been used in such different disciplines as anthropology and mathematics, and has been applied to both small and large social groups such as schools and work environments (Freeman, 2004). This type of analysis will often use mathematical models to explain the relationships existing among the members of a group, which are described as ties relating individual actors represented as nodes.

As for many other social phenomena, a good example being the spread of virtual tribes or the communities (Cova & Kozinets, 2007), Internet has been a powerful accelerator for an already existing and well-rooted phenomenon. It acts as a simplifier of relationships among members of a group, and eliminates existing hierarchic levels and, obviously, physical distance. At the same time, the virtual nature of the web encourages interpersonal relationships that are less involving from the psychological point of view and are less dangerous (Wallace, 2001). The user's generated logic of Web 2.0 has further implemented the relational aspects of Internet. Indeed, besides the creative role that users apply to create different content away from professional roles and platforms, as for instance uploads of videos or photos or the creation of a wiki, they also have the possibility of building and shaping their relationships through social networks.

Lately, researchers have focused on the structure and working of online social networks employed to foster connection-sharing, social capital generation and effective communication (boyd & Ellison, 2007; Donath, 2007; Ellison et al., 2007).

The use of social networking is typical of the younger web surfers, the so-called digital native (also known as the Net Generation or Millennials), the cohorts of young adults who have grown up with personal computers, cell phones and the internet (Bennet et al, 2008, Prensky, 2005). But these constantly connected individuals – always in touch any time and any place – are not the only subscribers to social network sites. Indeed, even if as some research highlights (Pew Internet, 2003) the digital divide still separates younger from not so young people, recent data underline that the age of social network users is increasing. User statistics of Myspace, and especially Facebook, demonstrate that the more visible expansion is among users aged 26 or more. In the US (Inside Facebook, 2009a) Facebook has been growing particularly quickly amongst people over 45, and even though the fastest growing age group is still 26-34, the number of women over 55 has grown 175.3% since the end of September 2008. The trend has since been confirmed: in the first half of 2009 Facebook's most substantial increase was amongst users 35-65 years old (Inside Facebook, 2009b). With the rapid growth among older users, the majority of US Facebook users are now over 25 years old. At the beginning of March 2009 there were 6 million Facebook users aged 13-17, 19.5 million aged 18-25, 13.4 million aged 26-34, 9.7 million aged 35-44, 4.6 million aged 45-54, and 2.8 million over 55 (Inside Facebook, 2009c).

It is very clear from the statistics that the percentage of users in older age groups is still growing (Table 1).

Age	December 2009	March 2009
13-17	10 %	11%
18-25	35 %	29%
26-34	24%	23%
35-44	19%	17%
45-54	12%	8%
55-65	7%	5%

Table 1: Age groups of US Facebook's users (Source: table assembled from statistics taken from Inside Facebook www.insidefacebook.com)

The ageing trend is related to Facebook's increasing success: according to Nielsen data (Lucchini, Vignale 2009) the number of registered users went up by 168% between December 2007 and December 2008, thus allowing Facebook to outgrow Myspace and become the world's most popular social network (Bains, 2009).

Most of the available research concerns US users of social networks; therefore the aim of the following sections is to refer to recent Italian research to determine whether there are distinct differences between Italian social network users which can be attributed to age.

The work aims to check whether the following variables can identify different groups of users: a. level of technological skills (i.e. being a digital native or not); b. way of accessing Internet (occasional access versus continuous online connection); c. motivation that drives people to open an account on a social network (i.e. searching for old school-mates or contacting someone you met the night before).

2 The Research

Explorative research was conducted on quantitative data collected through a participative research on social networks promoted by an Italian research company, SWG (DiarioAperto 2009).

The data collection window of the quantitative on-line questionnaires was from March 2009 through May 2009. To prepare the questionnaire, the researchers collected impressions and ideas from opinion leaders, i.e. blog users and Internet surfers, about the topics related to social network use that they considered to be most interesting.

The questionnaire was then circulated by uploading it onto the main social networks and onto the websites of partner institutions such as Sole24Ore.com and Studenti.it. The data were collected using a computer-aided web interview method without applying stratification.

2.1 The Sample

A sample of 1,251 complete returned questionnaires was used, which is obviously a non-probabilistic sample and although it cannot statistically reflect all Italian social networks users, the results can certainly give an indication of the direction of current trends.

The sample is almost equally divided between gender: 55.10% male and 44.90% female. From the distribution among the age groups of the respondents it would appear that the progressive ageing of social network users has not yet spread to Italy (see Table 2 below). However, the method used to collect data was through social networks so responses were probably obtained predominantly from the most assiduous users.

	Tot	Male	Female
	1.251	690	561
18-24	395	199	196
	31.60%	28.90%	34.90%
25-34	446	240	206
	35.70%	34.80%	36.70%
35-44	263	157	106
	21.00%	22.80%	18.90%
45-54	111	70	42
	8.90%	10.10%	7.40%
55-65	24	15	8
	1.90%	2.20%	1.50%
>65	11	8	3
	0.90%	1.10%	0.60%

Table 2: Sample by age (Source: table assembled from SWG data)

2.2 Modality of Fruition

According to the literature the modality of access to Internet is associated with diverse types of users (Hargittai, Hinnant, 2005); again the following table (Table 3) reflects the modality of connection of the sample: almost all the respondents, regardless of age, declare they connect to the Internet at least once a day, although there is a slight preference for continuous connection for the younger age groups.

Among the different applications available, e-mail services are used by 99% of the sample, followed by instant messaging such as MSN or Skype, which is used by 82%. Chat facilities, like Facebook's chat, are used by 65% of the sample, 54% declare they take part in forums, 39% have subscribed to feed readers, 29% have subscribed to news groups, 26% use collaborative platforms, e.g. writing a wiki, and 24% have a twitter account. The results are already partially obsolete: indeed, according to Nielsen data (2009) social networks are now the fourth most popular online activity, ahead of personal e-mail.

Age groups	18-24	25-34	35-44	45-54	55-65	>65
All the time if possible	174	241	148	57	10	5
	44.10%	54.00%	56.20%	50.80%	41.40%	42.10%
Every day	199	183	111	49	12	6
	50.30%	41.00%	42.00%	44.30%	50.10%	52.60%
Three times a week	15	17	5	5	2	0
	3.70%	3.90%	1.80%	4.90%	8.60%	0.00%
At least once a week	5	3	0	0	0	0
	1.40%	0.80%	0.00%	0.00%	0.00%	0.00%
At least once a month	2	0	0	0	0	0
	0.40%	0.10%	0.00%	0.00%	0.00%	0.00%
Less than once a month	0	1	0	0	0	1
	0.00%	0.30%	0.00%	0.00%	0.00%	5.20%

Table 3 - How often are you connected? By age (Source: table assembled from SWG data)

In order to detect generational differences among social network users, the data regarding use of applications have been assigned to age groups. Table 4 shows the results.

Age groups	18-24	25-34	35-44	45-54	55-65	>65
Email	387	438	263	110	24	10
	97.80%	98.20%	100.00%	98.40%	100.00%	94.80%
IM	359	367	205	62	10	4
	91.00%	82.30%	77.90%	55.30%	40.20%	34.20%
Integrated chat	255	281	169	62	9	7
	64.50%	62.90%	64.20%	56.00%	36.50%	60.60%
Forum	189	219	151	50	13	1
	47.90%	49.20%	57.20%	44.50%	54.90%	5.20%
Feed Reader	104	131	99	29	4	1
	26.30%	29.30%	37.50%	26.20%	17.00%	10.50%
News-group	78	123	82	26	12	1
	19.80%	27.70%	31.00%	23.20%	48.80%	5.20%
Collaborative platform	76	85	64	20	4	1
	19.30%	19.00%	24.50%	17.80%	17.00%	5.20%
Twitter	40	70	60	13	3	1
	10.20%	15.60%	22.60%	11.40%	12.10%	5.20%

Table 4 - How many applications do you use? By age (Source: table assembled from SWG data)

Differences can be detected among the oldest users who appear to be more at home taking part in forums and newsgroups than using integrated chat options.

2.3 Social Networks and Modality of Fruition

Almost all of the respondents (98%) have subscribed at least to Facebook, which can therefore be considered the most popular social network in Italy. Almost half of the sample say they visit their favourite social network every day, 23.10% any time they can and 17.50% connect to it three times a week. Only around 10% of respondents declare a less frequent connection (5.40% at least once a week, 2.10% at least one a month and 2.70% less than once a month) indicating that visiting online social network sites is, for most of the sample, an everyday activity. Indeed, even among the oldest age groups the most frequent choice of connection is every day (64.60% for the 55-65 age group), while among the 35-44 year-olds "three times a week" is chosen by 23.40%, compared with 21.40 for "always if possible".

According to the literature (boyd & Ellison, 2007, Hargittai, 2007), the principal function of social networks is to help people to connect with people they already know. Indeed, unlike others kinds of communities still crowding the web, rather than being centred on the sharing of an interest, social networks are centred on the individual. Individuals are at the centre of their relationship network, i.e. they become the centre of their own community. This need to put themselves on show (Papacharissi, 2009), to become the node of a relational network is supported by the different forms of "being in a window" (Codeluppi, 2007) that have characterised the social development in the last ten years.

The available data confirm the idea that Facebook helps people to stay connected, just as its homepage claims. Respondents were asked to declare their degree of agreement or disagreement with some of the most obvious uses of a social network (Table 5).

	To stay connected with friends and acquaintances	To flirt	To meet new people	To share information	To organise an event or to support a cause	To do business or find business contacts	To promote yourself or your business
Strongly agree	731	20	53	455	301	70	104
	58,50%	1,60%	4,20%	36,40%	24,10%	5,60%	8,30%
Agree	376	92	254	551	469	151	223
	30,10%	7,40%	20,30%	44,10%	37,50%	12,10%	17,80%
Disagree	104	255	442	184	343	338	325
	8,30%	20,40%	35,40%	14,70%	27,40%	27,00%	26,00%

Strongly disagree	34	869	496	56	134	680	587
	2,70%	69,50%	39,70%	4,50%	10,70%	54,40%	46,90%
I don't know	6	14	5	4	5	12	12
	0,40%	1,10%	0,40%	0,30%	0,40%	1,00%	1,00%

Table 5 – What do you use your social network for?(Source: table from SWG data)

For the sample, the main use of social network is to stay connected with friends and acquaintances (a combined score of 88.60% for "agree" and "totally agree"): the second most popular use is to share information (80.5% total for "agree" and "totally agree). Similar percentages pertain for disagreement regarding the use of social networks to flirt (89.90% total for "disagree" and "strongly disagree") or to meet new people (75.10% total for "disagree" and "strongly disagree").

Looking at the differences among the age groups it can be seen that there is a slight but significant disagreement between 35-44 year olds and the 45-54 year olds regarding the first use of social network (10.80% compared with 18.00% disagree with the statement that social network are used to stay connected with friends). However, the same age groups are not so strongly in disagreement with the statement "I use my social network to meet new people". As far as the function of sharing information is concerned there is not distinction among age groups.

Other notable differences can be found concerning the use of social networks to do business or find business contacts. The older respondents show a slightly higher percentage of agreement that the younger ones (see Table 6).

The same higher percentage among older respondents also reflect the use of social networks to "promote yourself or your business".

Age groups	18-24	25-34	35-44	45-54	55-65	>65
Strongly agree	7	35	24	3	3	0
	1,70%	7,80%	8,90%	2,30%	11,00%	0,00%
Agree	42	59	34	14	3	0
	10,50%	13,20%	13,00%	12,20%	11,00%	0,00%
Disagree	98	133	77	23	6	1
	24,80%	29,90%	29,10%	21,10%	24,40%	5,20%
Strongly disagree	240	217	128	72	13	10
	60,70%	48,70%	48,80%	64,40%	53,60%	89,50%
I don't know	9	2	0	0	0	1
	2,30%	0,40%	0,20%	0,00%	0,00%	5,20%

Table 6 –Use of social networks to do business or find business contacts (by age) (Source: table assembled from SWG data)

3 Conclusive Remarks

Extant research (Ellison, Steinfield & Lampe, 2007) confirms that social networks, and particularly Facebook, are tightly integrated into the daily lives of their users. Regardless of age, respondents prefer to be continuously connected to their social network and users like updating their status and sharing thoughts and ideas with friends.

Unlike other internet applications, such as collaborative platforms or interest-based communities, social networks are centred on the relational side of the web (Armstrong, Hagel, 1996). User-friendliness, the possibility of handling different applications on a single platform, and the possibility of using the site as a window of yourself (Papacharissi, 2009) have decreed Facebook as the preferred social network worldwide, having outgrown the "older" Myspace (Bains, 2009) to become the world's most popular social network.

The news about the progressive aging of Facebook subscribers (Inside Facebook, 2009a, 2009b, 2009c), together with the availability of detailed quantitative research on social networks in Italy (Diario Aperto, 2009), have been the drivers of a generational approach to the use of social networks in Italy.

Indeed, the data give some ideas for such a generational approach: the older subscribers, for instance, seem to use Facebook even for work-based tasks and are less confident about using social networks to reconnect with friends. At the same time, a slightly higher percentage of older users continue to take part in forums and newsgroups that have been practically abandoned by younger users.

The aim of this work was to check whether groups of social network users could be differentiated by reference to levels of technological skills, ways of accessing the internet and motivation to join a social network. It can be concluded that groups of users can be differentiated by referring to their age group together with these variables.

The work has obvious limitations: the first, and most important one, concerns the available data. The questionnaires were collected without paying attention to any kind of stratification, thus the sample is statistically unrepresentative, with the older users representing a very small percentage of the total sample. Therefore any conclusions drawn from the data cannot be statistically relevant and will only offer suggestions about on-going trends.

Further research is clearly needed to obtain more relevant data so that what is happening in Italy can be compared to what is happening worldwide.

Bibliography

Armstrong, A. & Hagel, J., 1996. The Real Value of Online Communities. *Harvard Business Review,* 74(3), pp.134-41.

Bains, L., 2009. *Facebook Overtakes MySpace as Most Popular Social Networking Site.* January 27th. Available at: http://www.switched.com/2009/01/27/facebook-overtakes-myspace-as-most-popular-social-networking-sit [Accessed 10 December 2009].

Bennett, S., Maton, K. & Kervin, L., 2008. The 'Digital Natives' Debate: A Critical Review of the Evidence. *British Journal of Educational Technology,* 39(5), pp.775–786.

boyd, d. m. & Ellison, N.B., 2007. Social Network Sites: Definition. History and Scholarship. *Journal of Computer-Mediated Communication.* [Online], 13(1). Available at: http://jcmc.indiana.edu/vol13/issue1/boyd.ellison.html [Accessed 10 December 2009].

Codeluppi, V., 2007. *La Vetrinizzazone Sociale. Il Processo di Spettacolarizzazione degli Individui e della Società.* Torino: Bollati Boringhieri.

Cova, B. & Kozinets, R., 2007. *Consumer Tribes.* Amsterdam: Butterworth-Heinemann.

Donath, J., 2007. Signals in Social Supernets. *Journal of Computer-Mediated Communication.* [Online], 13(1). Available at: http://jcmc.indiana.edu/vol13/issue1/donath.html [Accessed 10 December 2009].

Durkheim, E., 1971. *La Divisione del Lavoro Sociale.* Milano: Edizioni di Comunità.

Ellison, N.B., Steinfield, C. & Lampe, C., 2007. The Benefits of Facebook 'Friends:' Social Capital and College Students' Use of Online Social Network Sites. *Journal of Computer-Mediated Communication,* 12(4). Available at: http://jcmc.indiana.edu/ vol12/issue4/ellison.html [Accessed 10 December 2009].

Freeman, L. C., 2004. *The Development of Social Network Analysis: A Study in the Sociology of Science.* Vancouver: Empirical Press.

Hargittai, E., 2007. Whose space? Differences among Users and non-Users of Social Network Sites, *Journal of Computer-Mediated Communication.* [Online], 13(1). Available at: http://jcmc.indiana.edu/vol13/issue1/hargittai.html [Accessed 10 December 2009].

Hargittai, E. & Hinnant, A., 2005. *New Dimension of Digital Divide: Differences in Young Adults' Use of the Internet.* Paper presented at the Eastern Sociological Society. March. Washington DC.

Inside Facebook, 2009a. *Fastest Growing Demographic on Facebook: Women Over 55.* February 2nd. Available at: http://www.insidefacebook.com/2009/02/02/fastest-growing-demographic-on-facebook-women-over-55/ [Accessed 10 December 2009].

Inside Facebook, 2009b. *Number of US Facebook Users Over 35 Nearly Doubles in Last 60 Days.* March 25th. Available at: http://www.insidefacebook.com/2009/03/25/number-of-us-facebook-users-over-35-nearly-doubles-in-last-60-days/ [Accessed 10 December 2009].

Inside Facebook, 2009c. *November Data on Facebook's US Growth by Age and Gender: Young Men Following the Women.* December 3rd. Available at: http://www.insidefacebook.com/2009/12/03/november-data-onfacebook%E2%80%99s-us-growth-by-age-and-gender-young-men-following-the-women/ [Accessed 10 December 2009].

Lucchini, E. & Vignale, M., 2009. Ho Messo la Rete nel Taschino. *Io Donna.* 3 May, p. 79.

Nielsen, 2009. *News Release: Social networks & Blogs now 4th most Popular Online Activity, ahead of Personal Email.* March 9th. Avalable at: http://giovannacosenza.files.wordpress.com/2009/03/ricerca-nielsen-sui-social-networks.pdf [Accessed 21 December 2009].

Papacharissi, Z., 2009. The Virtual Geographies of Social Networks: a Comparative Analysis of Facebook, LinkedIn and ASmallWorld. *New Media Society*, 11, pp. 199-220.

Pew Internet, 2003. *The Ever-Shifting Internet Population: A New Look at Internet Access and the Digital Divide.* Available at: http://www.pewinternet.org/Reports/2003/The-EverShifting-Internet-Population-A-new-look-at-Internet-access-and-the-digital-divide. [Accessed 10 December 2009].

Prensky, M., 2005. Listen to the Natives. *Educational Leadership,* 63(4), pp.8-13.

Scott. J., 1988. Social Network Analysis. *Sociology*, 22(1), pp.109-127.

Tönnies, F., 1979. *Comunità e Società.* Milano: Edizioni di Comunità.

Wallace, P.M., 2001. *The Psychology of the Internet.* Cambridge: Cambridge University Press.

Marco Centorrino

The Image of the "Digital Native" and the Generation Gap

1 Introduction

Rarely has the representation of the very young in Italy focused so strongly on negative images. They are viewed almost exclusively as deviant, excessively precocious, and difficult to integrate into traditional social structures. Young people have, however, played a decisive role in the dynamics of the country's modernisation, especially in recent years. We need merely think of the process of cultural industrialisation in 20th century Italy, which was characterised by a series of anomalies which led to non-linear development and, at the same time, created profound differences with other European countries and the Western world as a whole[1]. The situation changed radically in the last decade of the century, with the advent of the so-called digital revolution which resulted in a sort of global 'reset'[2]. In fact, studies of cultural consumption show that in the 1990s a virtuous mechanism was set in motion on a national level, whose propulsive force was an increasingly interested and enthusiastic public[3]. This trend would be consolidated with the passing of time (Censis-Ucsi, 2006).

These sudden changes have a significant generational aspect. The young, above all, have made communication "the distinctive code of social life" (Morcellini, 2008, p. 11), allowing the growth of more innovative tools and

---

1   In outlining its various evolutionary phases, David Forgacs (1990) and Fausto Colombo (1998) describe a process characterised by "fits and starts", in which rapid acceleration is countered by suddenly *slamming on the brakes*. This process led both to the underdevelopment of some sectors and the absolute originality of others compared to the rest of the world. For further details, see also Morcellini (2005).
2   In some ways, the culture industries of the various countries were forced to radically reorganise in order to cope with new market dynamics. In some cases, including Italy, there was an opportunity to make up for lost time.
3   Without going into detail, we should remember the fact that in early 2000, Italy became the European leader in the use of satellite TV (Davies, 2002), with a penetration rate of 27% (this figure represents the sum of those with normal subscriptions – 1.9m families – and the sector of the public which accessed the service illegally – over 2.1m) and an average spend per user of €40 (a figure similar to the €45 of Great Britain, until that point the benchmark for the entire pay-TV market in Europe). Yet surprisingly, Italy was one of the last countries to establish a TV industry (50 years previously and only with the support of state investment) and did not see the spread of modern standards in terms of technology (colour broadcasting) and content (commercial TV) until the early 1980s.

limiting damage to those media that despite the general increase continue to record a negative trend[4] (Tirocchi, 2008).

Within the group of the new generations, moreover, we find a further core, which has a significant impact on cultural consumption: this core is composed of young people who were born when network technologies had already completed the first phase of national diffusion and are now aged between 12 and 16. Those youngsters, in other words, whom the media and public opinion view above all as 'tough kids'. As a group, their experience is completely different to that of their older brothers and sisters, because to put it simply, they arrived after the advent of internet and mobile phones. This difference is obviously accentuated when we compare them to adults (including their parents), who have gone through the 'rite of passage' from analogue to digital. Even before their predecessors, then, these kids jumped on the bandwagon and took the country towards post-modern life, thereby helping to accelerate the rate of development of Italian society.

Nevertheless, the media system – and more or less consequentially, public opinion itself – does not seem to have realised this evident fact[5].

My intention here is to provide empirical confirmation for these working assumptions and identify the possible causes of the phenomenon on the one hand, and on the other to establish the salient characteristics of a generation whose qualities still seem to elude us.

2 Tough Kids and Videophones

The study I performed[6] on some of Italy's main dailies and periodicals[7], although not comprehensive in nature and lacking techniques for content analy-

---

4 In the sector of daily newspapers, consumption by young people is well below the overall threshold, even though it is more in line with average consumption in terms of weekly and monthly publications.
5 The situation was vividly described by Alfredo Milanaccio: "In late autumn 2006, and for a number of weeks, dailies and weeklies, news programmes and talk-shows [...] raised public awareness (?!) of the fact that a new, dangerous, transgressive and derisive species of aliens had descended upon us. Since the species was, in fact, alien, it was provisionally given a fairly generic name: the young. Moreover, these youths, these aliens, not only proved to be derisive and transgressive, but were also dangerous, very dangerous even, because almost all of them, almost always, went around armed. Not with firearms, or knives, but with a new kind of weapons – ultra-technological toys – that they knew how to use with incomprehensible and disturbing skill, and to which they had given strange names: Videophone, YouTube, E-mule, SeeMe Tv, MySpace, Weblog, Internet and others." (2007, p. 3).

sis, allowed me to assess which topics are associated with the representation of 12 to 16-year-olds. In other words, I was interested in those cases when this target crosses the threshold of newsworthiness, sometimes all too easily.

The main theme seems to revolve around *bullismo* (bullying), a term which covers acts of prevarication and/or violence committed by groups of young people or individuals. The term is related to the Italian words *bulli* and *bulle* (tough guys and girls, bullies) and was hardly heard at all until a few years back, but the use of these expressions now is quantitatively significant. One just has to look at *Il Corriere della Sera: in 2001 the newspaper* used the word 'bullismo' in 21 articles while the number rose to 242 in 2007. *La Repubblica*, meanwhile, went from 68 articles in 2004 to 108 in 2005, 325 in 2006, and reached 507 in 2007; *La Stampa* went from 109 in 2005 to 229 in 2006 and then to 500 in 2007[8].

In all these cases of the references to bullying the situation is represented as a problem, with little space dedicated to the recounting of factual evidence and much more space given over to the comments of journalists and experts. The views of the former are almost always apocalyptic in tone:

> [...] there is no doubt that things have got worse, that violence has become our daily bread, made with the devil's flour. At high school, kids arrive dressed like hooligans, with a surly air, a bold swagger, a cold stare. They have already sat through millions of hours of cartoons revolving around fights, hand grenades, bursts of machine-gun fire, no-holds-barred struggles for supremacy. They are sponges dripping with dirty water [...]. Compared to twenty years ago, these youngsters have a burnt-out nervous system [...][9]. (Lodoli, M., 2008. Lo scalpo. *la Repubblica*, 7 May)

The reports make no gender distinctions: girls and boys are lumped together in a single vision.

As far as the experts' comments are concerned, they tend to simply provide advice and suggestions.

> "School on its own is not enough. It's the families who should punish their children in order to educate them. But nowadays unfortunately parents don't hand out punishments anymore and the little tyrants rule the roost at home and school" (Gasperetti, M., 2008. Prendi troppi ottimo: e gli mettono la testa nel water [10]. *Il Corriere della Sera*, 20 Jan.)

---

6   For further details on the methods adopted and for a more extensive discussion of the results, see Centorrino (2008).
7   La Repubblica, Il Corriere della Sera, La Stampa, Il Giornale, Panorama (all studied in print and online versions, between December 2007 and June 2008).
8   Data collected by consulting the online archives of three Italian dailies.
9   The translation of quotations from newspapers are mine.
10 Interview with Paolo Fuligni, psychologist and psychotherapist, expert in problems related to bullying.

Basically, the picture drawn by the media leaves no room for alternative opinions. Each episode of news involving young people is framed within the context of a crisis of values, violence and malice. The idea that bullying can be used as a wide-angle lens through which to observe all these events is hardly ever questioned. On the contrary, the Italian press considers this to be a 'growing trend'. We are presented with that scenario which Elisabeth Noelle-Neumann (1980) summarised in her model of the "spiral of silence": the realisation that a single, homogeneous, stereotyped representation is used for much more complex phenomena.

The second way of categorising young Italians involves the concept of *precocity*, associated with any behaviour, which is not considered appropriate for the age of the social actors. This term also covers another series of attitudes, whose negative potential is accentuated by the fact that the protagonists involved are so young.

If we break down this thematic group, which is also significantly large, into its basic elements, we find two main topics: sex and the consumption of alcohol/drugs.

In the case of the topic of sex, journalists concentrate, on one hand, on discovering examples to report and, on the other, on attributing responsibility (replacing the advice from experts we encountered in reports on bullying). Pride of place is thus given to studies and direct accounts:

> [...] at 13 you're not that young anymore; some have already had sex. At 15, it's pretty much the norm for almost everyone. They disappear into the toilets and when they come out they've got an enigmatic expression. They've done it. (Marocco, T., 2008. Festini a luci rosse alle medie. E nei bagni...[11]. *Panorama*, 27 Mar.)

At the same time, these generational portraits often associate sexual precocity with young people's use of technology, which is seen as the root of all other behaviour. Mobile phones become 'pornophones', young people are described as being 'internet and mobile-phone addicts' and as needing showcases 'like a television audience'. But reports also focus on the behaviour of parents and models of reference. Let's see, for example, how *la Repubblica* comments on the elopement of a twelve-year-old with a twenty-seven-year-old near Palermo:

> [...] the picture that emerged shocked them [*the Carabinieri*]. Because not only the two families, but everyone, or almost everyone in the town, knew about the affair between the girl and the baker, and nobody was particularly worried about it, starting with Valentina's relatives. Even after her disappearance, they did not seem particularly worried about the girl's fate, once they had been reassured that nobody had kidnapped her to do her harm. This may sound like an unbelievable story, but unfortunately there are many others like Valentina, a little girl who grew up too fast without a mother to guide her. (Isman, G. &

---

11   Interview with Costanza, aged 15.

Ziniti, A., 2008. La bambina e il panettiere sposato, fuga d'amore shock a Monreale. *la Repubblica Palermo*, 5 June)

Precocious behaviour – as said – also includes the consumption of alcohol and drugs by the very young. Also here, the structure of media accounts does not significantly differ from the pattern seen above. The discovery of the phenomenon, however, depends above all on statistics derived from studies.

> Alcohol use is endemic amongst youngsters. [...] This worrying picture, emerging from the cross-referenced reports of the Department of Public Health and the Italian Statistics Institute, was presented yesterday on the occasion of the Alcohol Prevention Day. The research figures clearly seem to show that we are dealing with a real "alcohol generation" composed of adolescents, some of them very young, [...]. A drink has become a tool of socialisation, which youngsters would give up only for a large prize or participation in a TV reality show. (not available, 2008. Ragazzi e alcol è allarme, bevono due su tre. *la Repubblica*, 18 Apr.)

There remains the need to establish responsibility, usually by quoting further studies, in which attention remains focused on the culture industry. Compared to the situation regarding sexuality, however, in this thematic subgroup we find more clearly alarmist tones, expressed directly by the authors of the articles or in the experts' comments.

These two categories, which are definitely the most significant quantitatively speaking, provide the defining traits of a generational portrait: *tough guys and girls* and *videophones*. This description, moreover, is also perpetuated in literature, in books adopting a pseudo-ethnoanthropological approach. In her introduction to *Ho 12 anni faccio la cubista, mi chiamano principessa*[12], Marida Lombardo Pijola, for example, writes:

> I collected the material directly from the accounts of five youngsters and transformed it into short stories, which I reorganised by using a first person narrator and a style that adjusted their elementary language in order to explain the concepts and amplify their importance. (2007, p. 16)

On most occasions, however, such texts are patched together by reworking material from blogs and forums, even though in such contexts what obviously emerge are not so much young people's 'voices', but the expressions of the identity they use in cyberspace. There are no checks to see how far this coincides with what actually happens in objective daily reality.

Almost to prove the fact that in the view of the mass media I studied, young people are 'naturally' oriented towards bullying and precocity, we have a third

---

12 Published by Bompiani in 2007. The title translates as "I'm 12 and I'm a go-go dancer; they call me Princess".

group, in which even *normal* events become newsworthy and extraordinary. Priority here is given above all to direct testimonies:

> School – not just bullies, but also heart-warming stories. Chiara, Elisa and the volunteers of the Milan Provincial UNICEF committee ("Lettere", 2008. *la Repubblica Milano*, 24 May)

Rather than showing excellence and positive models, however, the media representation of these figures seems above all aimed at creating *structured oppositions*[13], in which deviance assumes greater importance[14].

In this informative circle, the tough kids are arrested, suspended and handed back to their parents. They have sexual relations, and spend their evenings consuming alcohol and drugs. They then seem to disappear almost into nothingness, their image dissolves. Moreover, since they are minors, we do not know their names (at the very most we have their initials, or false names), or faces. They are practically abstract entities, part of a circuit whose single LED only lights up to signal deviant acts or phenomena. As the years pass, rather than growing up, they remain at their school desks or on the streets, with their arrogant attitude. These stories about the teenage condition create immobility, and provide the image of a generation that fails to evolve, but rather – according to a sort of domino effect – perpetuates and aggravates its mistakes[15].

3 The Possible Causes

The research results offer two main explanations for the picture I have illustrated and, more in general, for the distance we see today between the world of adults and that of young people. The first is a generation gap which is much more accentuated than previously, while the second is the greater visibility of certain deviant phenomena and at the same time the quest for this visibility seen

---

13 Inspired by the work of Levy-Strauss, this expression adopted by semiologists indicates, within texts, the production of confines and differences, often attributing a higher value to one of the two members of the pair.
14 Moreover, this seems to be the tendency of all the news reports, in which more importance is attributed to the crime than to the punishment given to the culprit.
15 The trend has been continuing for some years, as shown by the extensive study carried out by Simona Tirocchi (2008), whose significant findings, based on a survey which started in 2004, coincide with many of the results set forth here. Of 360 articles examined – the data refers to surveys made in 2004 and 2005 – for example, already in that period the role played by minors in situations of unease/deviance was above all that of author/protagonist (48.9%), rather than victim (13.6%).

in young people's attitudes. Let's examine the issue step by step, starting with the historiographical point of view.

3.1 The 'in-between Generation'

In order to clearly establish our chronological parameters, we need to make a simple, approximate calculation: today's 12 to 16-year-olds were born between 1991 and 1996. According to our processing of ISTAT data[16], the average age for women giving birth in that period (referring both to the first-born and to subsequent children) was 26-29 years. Their parents, then, belong to a generation which grew up in the 1960s and 70s[17], a very distinctive period for many reasons, as we shall briefly see.

The society in which the future mothers and fathers of bullies were born was undergoing radical change, whose effects would however only be felt subsequently. This was a society divided by two opposing forces: a two-sided country, precariously balanced between modernity, tradition and backwardness, as Menduni (1999) argues. These were the foundations for the development, amongst other things, of the youth and workers' movements which would come to centre stage in the following decade. The divisions were accelerated by domestic migratory flows, and also by an international context which reflected the division between two social models: Eastern bloc solidarity and community against capitalist, individualist Western society.

Theirs is an 'in-between generation': people born in this period were too young to fully experience the economic boom and the movements of 1968, and mostly have parents who were adolescents in the heterogeneous post-war years.

They however became the guinea-pigs of the post-1968 years, especially as far as regards the education system, which the current forty-somethings entered while it was undergoing a period of extensive experimentation. In 1973, for example, there was a mass influx of newly-appointed teachers and the launch of the first innovative projects, but in the meantime Italian schools continued to have to deal with explicit selection practices (which forced some pupils into repeating a school year), based on a traditional elitist mentality, which ended up frustrating the spirit of the reforms. The system was being asked to experiment, while what it needed above all was to put itself to the test so that it could deal with internal paradoxes and maintain a leading role in a society heading for a profound and traumatic break with the past.

---

16 Including both the first and second five-year period of the 1990s.
17 Naturally, the calculation does not take into consideration any gender difference, considering that, in any case, it is reasonable to circumscribe to a decade the age difference between fathers and mothers.

So when the period of protest made way for that of subversion, many of the current parents of tough kids were not yet mature enough to fully espouse a political ideology or get actively involved in the student movements of 1977. They became so in the 1980s, a period in which, however, the "in-between generation" found itself having to rely on shaky points of reference, in some cases fictitious[18]. While the value of the brand-name and fashion triumphed, and points of reference were found in new forms of 'collectivism'[19](*paninari*[20], heavy metal freaks, etc.), today's forty-somethings had to deal with an ideological crisis. Paradoxically, then, on one hand they were forced to reformulate the cultural models based on strong political values which had been adopted by the older youths in 1968 and 1977. On the other, they failed to affect the dynamics of cultural change in the same way as the 20-year-olds had done in the 60s and 70s[21], since the values of capitalist consumerism extended to the whole of society, and its attendant attractions clearly reduced interest in politics and religion. But, above all, society felt the need for a "restoration" which could help resolve previous tensions, and this temporarily overshadowed youth movements.

Lastly, they encountered a further phase of the dispersion-recomposition of interpersonal and institutional relations in adulthood, following the 'collapse' of a system distinguished by the crisis of the welfare state and punctuated by the tumultuous sequence of events which started in the early 1990s[22]. To view it in socio-semiotic terms, following the model designed by de Saussure , they found

---

18 For example, they achieved economic independence in a highly contradictory scenario: from the 'mini-boom' of 1980 to the crises of 1982-3, from the clear recovery of 1984 to the crash of the Milan Stock Exchange in 1987.
19 In fact, the term assumes a completely different meaning from that with which it was associated in 1968, when there was talk of "collective action" to define the protest movements.
20 A youth subculture whose adherents wore designer clothes and frequented sandwich bars (*paninerie*).
21 A posteriori, previous studies provide us with two ways of interpreting the causes of the phenomenon, both summarised by Ginsborg (1998). According to the pessimistic view, the culture of Italian society has always been based on essentially negative civic values, characterised by individualism and a lack of solidarity and mutual trust. In this view, the movements of the 1960s and '70s were merely a parenthesis, while in the 1980s we witnessed a return to normality.
   At the same time, however, we can espouse a second view, arguing that Italian society in the 1980s evolved in two different directions, without either managing to prevail over the other.
22 From the Gladio scandal (1990) to the terrorist attacks in which the Antimafia investigating judges Falcone and Borsellino lost their lives ('92); from the first arrival of Albanian refugees ('91) to the 'Clean Hands' political corruption investigations which, since '92, have involved the leading names in politics and the economy; from the first Gulf War ('91) to the crisis of former Yugoslavia ('93), with the direct intervention of the Italian army on both occasions.

themselves facing the disappearance of *signifiers* and the need to establish conventional relations between new signs and symbols and the respective *signifieds*. In a nutshell, they witnessed (and took part in) a significant change in codes.

The children of these parents who had grown up in the 60s and 70s, however, are "digital natives" (Palfrey and Gasser, 2008): they live in a society in which the technological paradigm represents a universal point of reference. They have not been forced, at least so far, to deal with radical 'social upheavals', nor have their cognitive and assessment models had to operate within those comparative systems which their parents continue to refer to. An indicative example is the current interpretation of online reality: for adults it is a second dimension, seen in opposition to the traditional spaces of daily existence they have inhabited for most of their life. Those from the digital native generation, meanwhile, have a joint vision of the two dimensions, and do not have any terms of comparison. They thus move with greater ease in the spaces of contemporary life, and have left the older generation behind, on the other side of the "digital divide", to quote Furio Honsell's observation, taken up by Francesco Pira and Vincenzo Marrali (2007).

From daily practices to value universes, the distances were further widened by significant changes at the end of the 20th century. On the one hand, then, we have an 'in-between generation', which grew up in a context of temporariness, as I have tried to highlight. On the other, we have cognitively and behaviourally different generations, faced with different challenges and situations. These were first to be born in an era already characterised by the internet, at a time in history when they were called on to interpret as best they could the need to become an active part in the construction and deconstruction of reality. This was a task their parents never had to deal with (due to the events of the 1980s) or, at most, are only having to deal with now after, however, having gone through the rite of passage from analogue to digital.

3.2 Visibility as a Primary Need

The differences to which we just referred are confirmed above all in academic attempts to establish a suitable categorisation for the 'digital native generation', obviously going beyond the stereotypes of bullying and precocity. David Buckingham (2000), for example, sees them as an electronic generation, as opposed to those forecasts which see the media practices of the new generations as embodying the definitive decline of the idea of infancy. Vanni Codeluppi[23],

---

23 Interview quoted in Coppola, P., 2008. Tutti web, mamma e iPod: è la generazione senza nome. la Repubblica on-line, 07 Jul..

meanwhile, in the light of contemporary cultural consumption, talks of the "MTV generation" and "Lost generation"[24]. The psychoanalyst Gustavo Pietropolli Charmet (2008) provides a detailed analysis, claiming that we may rightly talk of "generation N" (where N stands for Narcissus) to emphasise the creativity and fragility of today's young people.

These are elements I found in the results of a further phase of my research, performed by means of participatory observation, starting in the forums linked to personal, collective and thematic blogs and continuing in more restrictive chatrooms[25] and social networks[26]. I started from the premise that above all the net is the main observation point exploited by the media and, more generally, by adults, for interpreting the digital native generation. In addition, however, I also had the opportunity to ascertain how these characteristics are manifested and observe the resulting phenomena. Summarising the results[27], the leitmotiv characterising the *youngsters'* behaviour seems to be a sort of 'fear of invisibility'. They crave to be seen, noticed and looked at. Emerging from the anonymous almost seems to be an obsession of the new generations. Leaving aside motivations of a psychological nature – an interesting topic for future studies – I found deeply rational explanations for all this. The social capital accumulated on the web in fact is transferred into daily life:

> Sometimes I find people recognise me when I go to parties: "Are you Aldebaran? Your blog's great, I often look at it". It makes it so much easier to "pull". But there are some real losers who just sit there all evening, nobody takes any notice of them. They don't make an impression, they're a waste of space; they don't have any friends because they're not interesting. (San, aged 12)

Accumulating capital also means managing to exploit the large numbers on which the internet is based. Managing a blog which attracts a large number of visitors, for example, also means earning through advertising banners or having a large target audience for e-commerce activities.

---

[24] The reference is to the well-known TV series, especially popular with the young, who have transformed it into a transversal phenomenon and set up discussion groups on the net.
[25] Such as Msn.
[26] Facebook, MySpace and Second Life. In total, I managed to collect around 120 interviews. Moreover, I had privileged access to a world in which, even without interacting directly with my interlocutors, I could easily observe others' relations. Many of the tools used (especially the social networks) in fact allowed me to see how the discussions evolved, such as inside a focus group. All I had to do was provide input, then step back and see how the discussion developed, by means of posts, messages or text strings.
[27] Here too, for further details on the methods adopted and for a more extensive discussion of the results, see Centorrino (2008).

The production of the content of a blog unconsciously seems to be inspired by what amount to marketing strategies: 'a marketing of self representation', in which the key approach is to amaze by means of transgression[28]. This means exploiting the expectations of others – those of the adult generations, more than those of the young – to accede, via the web, to the obscene, to the Goffmanian backstage of others' lives.

This is connected to a further aspect which seems to accentuate the generation gap. Some deviant behaviour, whether real or artfully constructed is in fact common to many generations but today we are perhaps for the first time seeing it so blatantly, intentionally and defiantly presented. There are various actions – boasting in a blog that you have forged your father's signature on a letter justifying absence from school; filming yourself when you're drunk and uploading the video on YouTube; getting original comments about you written in the class register and then documenting the fact by taking a photo and posting it on the net[29] which, according to traditional dichotomous logic of transgression, must be shown to the peer group to achieve status, but that at the same time must be hidden from others since they expose one to the risk of punishment by adults. This is a danger that many of the young people interviewed did not perceive. In sociological terms, we can see the phenomenon not so much as the result of a crisis of the individual institutions (school, family, etc), but above all as a mechanism which leads to the distribution of what Weber's defines as "charismatic power". We have a short-circuit of *authority*, behind which there lies a vacuum of *authoritativeness*, a value in fact belonging to those who hold charismatic power, which allows them to exercise *authority*. In reality, these mechanisms are fully functional within the peer group, while we find their absence in other contexts (for example, in the relations between adults and teenagers).

The same media usage model with the youngest being involved above all in the guise of producers and the adults as consumers overturns the usual hierarchies of the transmission of knowledge i.e. from the oldest to the youngest. We are witnessing a scenario in which the new generations are deprived of the traditional polar values of good, i.e. what is right, and evil, i.e. what is wrong. Lacking such points of reference, the social actor is bound to seek refuge in the

---

28 Asking young people for suggestions on how to create a successful blog, I managed to highlight a series of key elements. "Put interesting stories in. Grown-up stuff. Like you sniffed coke and then set fire to two niggers (Stick, aged 15)"; "you have to talk about things that others don't know and show you're able to do them. How to roll a joint, how to pull at the disco, what to do if you get a girl pregnant (Gary, aged 14)".

29 By collecting material on the net, Rizzoli has published two books on this topic, edited by an author with the pseudonym John Beer: "La classe fa la Ola mentre spiego" (2006) and "L'alunno è stato assente causa assedio testimoni di Geova" (2007).

search for consensus[30], which can be translated into a desire to appear (and conversely into a "fear of invisibility") and be recognised.

At the same time, a kind of transgression that is expressed in a single approach (showing to everyone, hiding from no one) and thus flaunted is much more visible. It significantly conditions our image of 12 to 16-year-olds. This evident manifestation means that we do not feel as if we are spying, but in reality it is consciously presented to us using new technologies, which act as an indispensable interface in the certification of transgressive actions.

4 Conclusion: the Hidden Characteristics of the 'Digital Native Generation'

The generation gap and the consequent ways in which 'digital natives' are represented actually in my opinion end up obscuring a series of abilities possessed by youngsters which will be a fundamental resource for the future evolution of Italian society. In conclusion, I will try to reconstruct – on the basis of the data collected – an identikit of 12 to 16-year olds which goes beyond the currently dominant images.

First of all, these 'digital natives' are multi-taskers: in other words they can process a whole range of information at the same time. They are able, for example, to talk on their mobile while surfing the net with various windows open, chatting and listening to music in the background. They are, consequently, multimedia beings: not in the sense given to the term up until the 1980s, connoting the potential and ability to use various means of communication, but rather in the sense that they are able to exploit various media at the same time. This also leads to convergence, to the extent that they are able to integrate – and encourage the integration of – apparently unrelated technologies. They can, for example, establish communication between cameras and televisions, mobile phones and satellite navigators, musical instruments and computers.

Moreover, these 'digital natives' are extremely selective in terms of media content: they are unfamiliar with the *push* model, in which information is pushed towards a mass public by a limited number of broadcasters, while they

---

30 This aspect is widely dealt with in studies of moral and political philosophy. The concept was originally introduced within utilitarian theory (originating with writers such as Hume, Bentham and Mill) whereby in contemporary society increasing importance is given to the idea of "public ethics" based on consensus achieved by using arguments which tend to pursue the principle of utility, or greater wellbeing.
In politological terms, Perelman's contribution is also relevant, and maintains that a new form of rhetoric revolves around the idea of consensus. On this topic, apart from the well-known Traité de l'argumentation: La nouvelle rhétorique. (1958), written by the author of Polish origin, see also Furnari Luvarà (1995).

are comfortable with *pull* dynamics whereby a single recipient can choose the messages that satisfy him most from a mass of broadcasters (consider preschool children, who are already able to navigate the articulated system of digital TV and find their favourite cartoon within seconds).

Furthermore, 'digital natives' are highly theme-oriented. In other words, they are almost completely unaware of generalism, and increasingly search for specialist knowledge (as reflected in the success of thematic channels or sites), which will allow them to reduce social "complexity" not by means of parcelled knowledge, but by exploiting a range of cultural specialisations.

Members of the 'digital native generation' are also multi-networked, and act in networks of social relationships they establish not only in daily life but also in the sphere of virtual reality. They are also clearly able to pass between the two levels. This continuous, constant activity makes them the principal actors in the processes of permanent communication.

Lastly, they demonstrate marked entrepreneurial abilities, exploiting all these tools in a context – that of the vastness of the internet – in which they demonstrate their skill in moving with great agility, combining internet interaction with everyday life: for youngsters, the screen, whether it be that of a videophone, a PC or a television, has ceased to be a *window on the world* and has become *part of the world*. All this has taken place while many studies are still wondering about the replacement of the original with the reproduction; the disappearance of the simulacrum and the contemporaneous affirmation of its copies; the true and the likely; or the risk that the individual may be alienated from traditional society and remain "trapped" in the net.

Members of the 'digital native generation' are basically active protagonists of the processes of construction and deconstruction of reality, able to exploit new practices of negotiation combining a tendency towards individualism and the collectivism typical of the networked society, and to generate "a relational wave effect" (Morcellini and Mazza, 2008) in which communication is a variable which plays a decisive role in overcoming individualist dynamics.

Bibliography

Buckingham, D., 2000. *After the Death of Childhood: Growing up in the Age of Electronic Media.* Cambridge: Polity Press.

Censis-Ucsi, 2006. *Quinto Rapporto sulla Comunicazione in Italia, 2001-2005: cinque Anni di Evoluzione e Rivoluzione nell'Uso dei Media.* Milano: Franco Angeli.

Centorrino, M., 2008. *Bulli, Pupe e Videofonini.* Acireale-Roma: Bonanno.

Colombo, F., 1998. *La Cultura Sottile: Media e Industria Culturale in Italia dall'Ottocento agli Anni Novanta*. Milano: Bompiani.

Davies, P., 2002. Pirates Have "two out of three" Pay-TV Homes in Italy. *New Media Markets,* 20(27), p. 5.

Forgacs, D., 1990. *Italian Culture in the Industrial Era, 1880-1980*. Manchester: Manchester University Press.

Furnari Luvarà, G., 1995. *La Logica del Preferibile: Chaïm Perelman e la "Nuova Retorica"*. Soveria Mannelli: Rubbettino.

Ginsborg, P., 1998. *L'Italia del Tempo Presente.* Torino: Einaudi.

Menduni, E., 1999. *L'Autostrada del Sole*. Bologna: Il Mulino.

Milanaccio, A., 2007. Presentazione. *Quaderni di sociologia: giovani e nuovi media*, 44, pp. 3-8.

Morcellini, M., ed. 2005. *Il Mediaevo italiano: industria culturale, tv e tecnologie tra XX e XXI secolo*. Roma: Carocci.

Morcellini, M., 2008. Leggere i segni e le provocazioni della new age. Senza pregiudizi ma anche senza abdicazioni. In: S. Tirocchi, 2008. *Ragazzi fuori: bullismo e altri percorsi devianti tra scuola e spettacolarizzazione mediale*. Milano: Franco Angeli.

Morcellini, M. & Mazza, B., ed., 2008. *Oltre l'individualismo. Comunicazione, nuovi diritti e capitale sociale*. Milano: Franco Angeli.

Noelle-Neumann, E., 1980. *Die Schweigespirale: Offentliche Meinung, unsere soziale Haut*. München: Piper Verlag.

Palfrey, J. & Gasser, U., 2008. *Born Digital: Understanding the First Generation of Digital Natives*. New York: Basic Books

Pietropolli Charmet, G., 2008. *Fragile e spavaldo: ritratto dell'adolescente di oggi*. Roma-Bari: Laterza.

Pira, F. & Marrali, V., 2007. *Infanzia, media e nuove tecnologie: strumenti, paure e certezze*. Milano: Franco Angeli.

Tirocchi, S., 2008. Ragazzi fuori: bullismo e altri percorsi devianti tra scuola e spettacolarizzazione mediale. Milano: Franco Angeli.

Matteo Treleani

The Access to Memory in Video Archives On-Line. Generational Roles on YouTube and Ina.fr

1 Introduction

The Internet can be seen as an audiovisual resource that, in the view of researchers, constitutes a 'real' collective memory. If we consider the role of younger generations in the use of new media and the significant influence these new forms of knowledge have on young people, it is useful to examine the ways in which the web addresses these users and gives them access to this audiovisual memory.

There are two websites which demonstrate a different approach to this question: Ina.fr, the online database of *Institut National de l'Audiovisuel*, a French state-governed commercial organisation, and YouTube, the largest source of videos on the web. 'Surfing' these archives is akin to 'surfing' the collective memory. The intention to build an audiovisual memory online has been explicitly put forth by INA (whose slogan until 2008 was "we build the future of your memory") and the same idea is implicit in YouTube, which can be surmised by the fact that it functions as a time capsule with a potentially infinite quantity of television programmes archived by users.

2 The Semiotic Perspective

The methodological perspective employed in this study is that of a semiotic comparative analysis, based on the effects of meaning that surfing the web engenders. The analysis proceeds through the study of two semiotic notions.

The first notion – that of the model user – is a notion applied by Giovanna Cosenza (2008, p.134) to new media which has as its foundation the model reader theorised by Umberto Eco (1979, p.50). The website addresses a user and reaches her/him by adopting rhetorical strategies. These strategies are discursive: they are part of the website and do not depend on real, or empirical, users. More than an enactment of a typical or ideal user, the model user is thus a textual construction that we find through an analysis of the places that the site creates for her/him in allowing her/him to act. The second notion is that a website is considered an interface, and not a space to be filled with content[1] ,as suggested by the supermarket metaphor which can be easily applied to Ebay(Stockinger, 2005), for example. The idea of a site as a place does not take into account the effects of meaning that the platform engenders: the interface determines the content and vice versa. In the study of an online archive we will be interested in the effects of meaning precipitated by the relocation of content: the influence of the context on the signification of the content is thus the objective of this analysis. As far as methodology is concerned, interfaces will be studied through interpretative paths as theorised by François Rastier (Rastier, 2001). These are the routes of meaning that the site grants the model user. The notion of a path takes us away from the idea of a static meaning, subjacent to the media, and is closer to the dynamism of the use of new media, where interaction plays a fundamental role. The meaning reveals itself in the interpretative movement through a multiplicity of anchors ('anchors' here denoting the supports of the interpretation of the object). Now these anchors may be part of the published content, a video (as in the context of broadcasting), or rather be external to it, such as the comments that a website posts with a video. The choice of the term 'interpretative' underlines that in the choice of the path we always have a production of meaning: in other words, reading a comment or clicking on one link rather than another has an influence on the meaning of the visit. These anchors of interpretation are, furthermore, conceivable as forms of expression in semiotic terms (a hyperlink, for example).

3 The Context of INA: the Editorialization of Audiovisual Documents

First of all we need to explain the particular context of INA. Its mission is the preservation of French audiovisual heritage, i.e. television and radio archives. The INA archives were, however, created with the purpose of commercially exploiting television archives (that is, selling them to professionals). The exploitation of this audiovisual heritage being historically entrusted to the cinema (especially in France, where film Institutes considered as heritage only 'authorial' films to the exclusion of commercial ones, (Vernet, 2007)), at its birth, INA was far from considering the aesthetic value of its television documents.

It is with the digitalization of this heritage, and with numerous more interesting possibilities offered by new technologies, that INA approaches the development of archives – aside from their preservation aimed at commercial gain. The plan for digital protection that consists of the mass digitalization of the heritage was signed in 1999 and INA has been officially responsible for the ex-

---

1 The notion of interface applied to new media we refer to Manovich, 2001, pp. 62-111.

ploitation of the audiovisual heritage[2] since 2000 its mission is to give access to the heritage to the public.

The publication of documents after their digitalization is thus the main operation of the Institute. INA relies on multimedia products for the editing of archive documents. After their indexation in a catalogue, certain documents are then re-published within specific projects of contextualization. They then become interactive DVDs, web archives or are even reassembled from archive films. These operations of re-contextualization are necessary for the exploitation of a large quantity of documents.

What interests us, from the point of view of the meaning of archived images, is that a system of editing that changes the context of publication changes the anchors on which the paths of interpretation of spectators rely. This change thus brings about an alteration in the meaning and value of published documents. In other words, if we see editing as a semiotic practice, where the operations of publication bring about expressive variations that imply consequent variations in the content, we can hypothesise that the editing of audiovisual documents has an influence on their meaning.

The comparison between Ina.fr and YouTube will allow us make clear some characteristics of these processes of re-contextualization in the case of audiovisual archives on the net. But first of all, we will look at the differences between the two sites from the point of view of both content and public.

4 YouTube

4.1 Contents

YouTube is a user-generated content site, i.e. videos are uploaded directly by users, although recent statistics indicate that only a small percentage of users actually generate content (between 0,5 and 1,5 %) (Manovich, 2008).

From a more specific perspective, the audiovisual genres proposed by YouTube are as follows:

The re-diffusion of TV programmes – especially old or foreign programmes that are of interest because of 'hearsay' on the web.

20% of content is actually musical video clips which also comprises 74% of the 50 most searched words. YouTube diffuses a large quantity of copyright protected video content. According to a study of online video by CNC con-

---

2 Article 49 du loi du 1er Août, 2000.

ducted by QualiQuanti[3], streaming is not considered by most users to be an illegal activity. Video clips are also a way of listening to music online without paying for it (looking for a song on YouTube is certainly faster than downloading it). Furthermore, Google search results for a song normally list YouTube video clips first.

YouTube is conducive to a form of entertainment which sees users taking pleasure from the misfortunes of others – *Schadenfraude* (Grevais, 2007).

4.2 Public

YouTube is a social network, and as a social network it is directed at an active user with a personal account and his or her own playlists. YouTube therefore contributes to the transformation of the web into an immense hard disk where data is stored[4]. The user of social networks also takes charge of the indexation of videos, through so-called 'folksonomies' (according to Crepel, 2008, 45% of tags used on Flickr are used only once, indicating the low reliability of widespread use of the system). Folksonomies generate a system where the information is managed bottom-up rather than top-down[5]. Furthermore, YouTube's public is young – aged between 18 and 24 (ComScore, 2008) – and so-called heavy streamers (those who watch videos on YouTube for more than 3 hours per day) are also the very young.

5 Ina.fr

Ina.fr was initially a business-to-business site, dedicated to broadcasting professionals. In 2006, the institute decided to widen its audience to non-professionals. The idea was to propose a dynamic archive emerging from the dichotomy (and opposition) between news and archive (Amit, 2009). Ina.fr was founded in 2006 – thus prior to the purchase of YouTube by Google and just before the boom of online video.

---

3 Les Nouvelles Formes de la consommation des images : TNP, TVID, VoD, sites de partages, piraterie, analyse qualitative, QualiQuanti (Daniel Bô, Claire-Marie Lévêque, Alexandra Marisglia), Paris, CNC, november 2007.
4 Fauré, C., Dataware et infrastructure du Cloud Computing. In Stiegler, Giffard, Fauré, 2009, p. 222.
5 Stiegler, B., Du temps-carbone au temps-lumière. In Stiegler, Giffard, Fauré, 2009, pp. 90-93.

5.1 Content and the Public

INA is the French national repository of television and radio archives. Its public and content are thus different from YouTube. A third of its content is news; another third comprises documentaries and reportage, and the remaining third is dedicated to entertainment (television series and features). As opposed to YouTube, there are no limits to the length of videos and the site provides a video-on-demand section. Officially, 80% of INA's digitalised archives are accessable online free (Amblard, 2007) of charge.

The target of Ina.fr is adult and also passive since users do not generate content. Content is useful for researchers, students or historians. According to Alexa's demographic charts most users are between 25 and 44 years of age (data which is only partly reliable). A certain number of hypermedia services are integrated into the site. In particular, we can single out the interactive frescoes: *Charles de Gaulle, paroles publiques* or *Jalons*. With a re-contextualization operated by historians, *Charles de Gaulle*, for example, gives value to a large quantity of old television news about Charles de Gaulle, with comments and historical reconstructions which increase understanding of the historical period. Other playful multimedia products try to allow a high-level interaction with the public. An interesting example is the Berlin Remix Festival[6]. Organised in partnership with DailyMotion, the festival presents a series of archive videos from INA available for reassembly by users. Several prizes are awarded to the best re-edited films made by users. If the remix is – alongside the database – one of the principal forms of contemporary creation for Lev Manovich, (2008) Berlin Remix is potentially the apotheosis of both.

5.2 The New Ina.fr

INA published a new version of its archives online in June 2009. The site, called "Archives pour tous", became the current Ina.fr. It was a specific marketing strategy aimed at broadening its audience. The title, moreover, evoked a difference between the real archives and the limited version available to the general public. The perspective has been thus inverted: archives are first of all for everyone, and only in the second instance does a more detailed version emerge for professionals. It is both an archive and a new media, according to Roei Amit, (2009), chief editor at INA.

How did Ina.fr attempt to achieve this change? It used Web 2.0 methods, following the example of YouTube (Gervais, 2007). The new site allows users to have a personal account, to insert comments and to evaluate videos, create

---

6   http://www.dailymotion.com/sas/BerlinRemix.

reading lists, export videos onto other sites and blog and share them through social networks. Videos carrying the label "INA" now circulate freely on the Web and in particular on sites broadcasting traditional media (Lemonde.fr, Liberation.fr, Telerama.fr etc.). These media make use of videos from the past to address current problems, in the same way as Ina.fr does. Content type has also expanded since the recent addition of 200,000 old advertisements made available by Ina.fr online.

5.3 Main Differences

YouTube	Ina.fr
User Generated Content	National Archives
Autonomy	Editorial Activity
Heterogeneity	Filter and Selection
Illegal Contents	Institutional Contents
Music Videos	News
Bottom Up Memory	Institutionalized Memory
Age: 18-24	Age: 25-44

Table 1: Main differences between YouTube and Ina.fr (Source: the author)

To conclude this section, we can state that the main differences between the two sites derive generally from the mode of publication of content. YouTube is a site containing user-generated content whereas Ina.fr is a national archive. This implies a significant heterogeneity for YouTube and an active process of selection and filtering for Ina.fr. Consequently we have autonomy on the one hand and a daily editorial activity on the other. The most popular type of material on YouTube is the video clip whereas on Ina.fr it is the news. On the one hand, part of the content is illegal and on the other, the all content is explicitly institutional content. Thus we see institutionalized memory in Ina.fr and "democratic" memory in YouTube generated by users. As far as the age of the public is concerned, the brackets are slightly different: 18-24 for YouTube, 25-44 for INA: this is a reflection of the differences in content. It is now necessary to see how these differences are reflected in various forms of online navigation, and thus how two types of model user are conceived by different forms of access to content.

6 Navigation in the Archive

Audiovisual archives such as YouTube and Ina.fr determine the way we access documents since the user is generally driven by the editorial choices of the site[7]. The first three results of a search on Google attract 80% of clicks. The user therefore trusts the hierarchical organization presented by the search engine (although the PageRank system and filters used by Google are not neutral and are often also stochastic (Ertzscheid, et al., 2007)). William Uricchio has noted the continuity in the evolution from old media to new media, and in doing so he points out that the protocols of metadata and filters used by search engines are never neutral in their advancement of viewing habits (Uricchio, 2009). Furthermore, cognitive science studies show that users are often superficial and distracted. It is thus necessary to consider the website as a technological support where content is fixed (Bachimont, 2007).

We can see the archives as a network, where documents are interconnected as happens in a rhizome. By following the model of Eco's encyclopedia (1984) a journey around the archives plays the role of a journey into the collective memory. Now, this journey is determined by the shape of connections among kernels of the rhizome: or the way of moving from one document to another in the website.

6.1 A Model of Online Navigation

We are now going to propose a simple cognitive model where the anchors of interpretative paths modulate the perspective of model users. The archives are thus a network where several documents are interconnected through a Web interface. This network is not static and fixed for all. It is modulated by user-generated research. Now we can see every click on the screen to move from one document to another as an interpretant that puts the ensuing document into perspective.

Charles Sanders Peirce defines the interpretant as the mediating representation that allows the passage between the representation and the represented object. Any activity of understanding would thus be a translation between various signs. Interpreting, according to Peirce, "fulfils the office of an interpreter, who says that a foreigner says the same thing which he himself says" (Peirce, 2003, pp. 57-68).

To conceive of a kernel of the network as an interpretant is useful for understanding the production of meaning present in every passage. For example, if you click on Ina.Fr's 'most popular videos' link you are offered videos that you

---

7   Carou (2007) considers the statistics about VOD in France by QualiQuanti.

understand to have been seen already by many other users and this influences your perspective of the document. Editorial choices thus determine the way content is accessed and they also allow for meaning to be accrued. Furthermore, seen positively, these tools have an epistemic character: they affect our knowledge (De Souza, 2005, p.11).

Moreover, digitalization leads to a loss of material interpretative anchors which we were used to in graphic reason (Goody, 1979, cited in Bachimont, 2000). For example, we lose some spatial anchors as the material structure of the book with its pages, which leads to a loss of orientation (Bachimont, 2000). The solution for disorientation, according to Bruno Bachimont (2000), is to give to users meta tools so to have a global view of the document to retain. In other words, a solution to the new problems spurned by digital technology would thus be the implementation of plans of navigation so that users can find themselves. Congruously with this position, the online archives of INA enable the user to see his or her own position within the archives. INA has chosen to create flexible margins of operation for users. Rather than force users into already mapped-out roads, Ina.fr allows them choose their own paths. Ina.fr offers tools that we shall call meta-discursive, so that users can choose their interpretative paths and consciously harness the possibilities of online navigation.

The context is thus constantly manageable for the user according to his or her purposes and interests. Ina.fr does not aim at erasing the context to prevent it influencing the meaning of documents; it aims rather at allowing the users to manipulate the context. This has the purpose of bringing to light the contextualization and the constitution of the archives by enabling users to be conscious of the processes activated through archival work.

6.2 YouTube

Following this model, let us now look at the differences between navigation on YouTube and navigation on Ina.fr. The homepage of YouTube (Figure 1) allows the user to see a selection of videos: videos that are being watched at that very moment, videos that are recommended and 'the most popular'. This type of categorization, independent of content, is directed at users who do not have a precise objective, and who can be placed in the category of the *flaneur* (Cosenza, 2008, p.135). Rather than looking for something particular, the *flaneur* navigates the Web as a surfer. Serendipity is the principle behind any search: coming across something that one was not looking for. This navigation without objectives is a product of the behavioral composition of the other users (Cardon 2008). It is quantitative: other users index videos and thus actively determine the list of most-seen videos and videos being watched at that moment. Also in line with a quantitative system is the hierarchization of documents i.e. the ordering of user comments on the screen. The passage from one knot of the net to another is founded therefore on delegation to other users of the choice of content and subsequent categorization of these choices (tags are planned by the same users through so-called folksonomies, Crepel 2008). The system of folksonomies mirrors the constitution of a bottom-up archive, where categorization is not determined by experts but by users themselves. YouTube is a container that seems neutral and democratic: its graphic aspects (like that of Google) are stylistically simple and sober (a white background and an almost total absence of decorative elements), giving the impression of a neutral and impartial container whose only role is to provide an available space for users. In reality the indexation and filters aren't neutral and drive the user into a time-wasting net, the goal being to provoke the maximum number of Hits. Meaning is therefore 'dispersed' to a thematic level, as passage from document to document is not determined by content. The result is an expansion of that type of attention that Katherine Hayles has called "hyperattention", as opposed to "deep attention" (that which focuses on a well-defined object for a long period of time) (Hayles, 2007 cited in Giffard, 2009, pp. 184-185). YouTube also incites constant activity in the form of comment or evaluation. The model user of YouTube focuses on the immediate instant without spending time examining the past with a look at the future which always rests restricted to the limits of a contingent action (Georges, 2009). This user loses him or herself in the archive.

6.3 Ina.fr

The objective of Ina.fr (Figure 2) is instead to allow the user to find his own position in the archive. The site aims at the editing of content with a selection clearly devoted to an adult target user base. The first page presents some news items, fruit of editorial activity, with the purpose of introducing the events of the past to be used to interpret the present.

Videos previously viewed are always available in an expandable window. It is the function called "mon parcours". The user is thus led to reflect on the connections of his journey around the archive. "Mon parcours" allows him to preserve a memory and a trace of his own navigation. The new Ina.fr site also offers the Mediagraph research system – a search engine with cluster technology (Figures 3, 4). A keyed-in term brings up a flashed animation showing all the documents to which it is connected. By clicking on a video thumbnail the system shows us the connections to other similar videos both thematically and visually. A few clicks will bring up a true local and perspective map of the archive. Making the relationship between documents explicit is therefore a meta-discourse on the archive. If we consider the archive as a labyrinth where one loses oneself, the Ina.fr user is a Dedalus who discovers a way to go out into the dimension that transcends it Dedalus escapes from the labyrinth by flying away: in other terms, he discovers the third dimension. At the same way, the user sees his own position by putting the archive into perspective through a local map.

In conclusion, we consider the archive of Ina.fr as constituted by a net of interpretants (just as with YouTube); the passage between documents is determined by the editing of content. If interpretants on YouTube were therefore otherusers, on Ina.fr, the passage from one knot of the net to the other is generally determined by editorial choices. The site nevertheless gives the opportunity to put the same net into perspective through meta-discursive functions.

6.3.1. Two Parallel Forms of Navigation

It is now necessary to consider that with the objective of reaching a target of a more sizeable public, and in particular a younger public, Ina.fr adopted a web 2.0 style of navigation. We are not going to deal with a comparison between the old and the new versions of the Ina.fr site, although it is important to emphasise that the meta-discursive tools are always proposed and never imposed.

Simplifying Ina.fr allows two types of navigation: one based on the style of web 2.0 – like YouTube – and another parallel type that we can define as "conscious". Web 2.0 institutionalises the notion of a young model user, whose searches are often superficial. Ina.fr therefore puts the style of conscious navigation into second place: tools are only potentially present and partly hidden by new graphics. With the objective of opening up to a young user base, Ina.fr harnesses some of the disorienting strategies of YouTube.

7 Conclusions

In conclusion, the young model user, the one generally concerned with web 2.0, seems to be conceived of by online audiovisual archives as a user who is not able to search independently for videos, but rather looks passively at what is offered with the objective of simple relaxation. Ina.fr shows nevertheless that we can allow the user to build connections in autonomous and divergent ways, guaranteeing a local vision of the archive and therefore of his own position in it. Considering that a system of unfocused navigation has great influence, especially on young people, the role of institutions does not appear to guarantee a satisfactory form of navigation or quality of content: rather, it provides critical apparatus that enables users to independently develop consciousness of their own navigation.

Bibliography

Amblard, M-C., 2007. Premiers enseignements d'une nouvelle offre: Ina.fr et Ina.media. *Archimages 07*. Paris: Institut national du patrimoine.

Allard, L., 2005. Express Yourself 2.0!.In: E.Maigret & E. Macé. *Penser les médiacultures. Nouvelles pratiques et nouvelles approches de la représentation du monde,* Paris: Armand Colin/INA.

Amit, R., 2009. L'Ina.fr est un Nouveau Type de Média. *L'Avenir de l'audiovisuel passe-t-il par le web? Dossier de l'audiovisuel*, Bry sur Marne, INA, Available at: http://www.ina-entreprise.com/observatoire-medias/dossiers/avenir-av-web/article-7.html [Accessed 14 December 2009].

Bachimont, B., 2000. L'Intelligence Artificielle comme écriture Dynamique : de la Raison Graphique à la Raison Computationnelle. In: Petitot, Fabbri ed. 2000, *Au nom du sens. Autour de l'oeuvre de Umberto Eco*, Paris: Grasset, pp. 290-319.

Bachimont, B., 2007. *Ingénierie des Connaissances et des Contenus*, Paris: Hermès.

Bachimont, B., 2008. La Conservation du Patrimoine Numérique: Enjeux et Tendance. *Patrimoine Numérique: Mémoire Virtuelle ou Mémoire Commune? Dossier de l'Audiovisuel,* Bry sur Marne, INA. Available at: http://www.INA-entreprise.com/observatoire-medias/dossiers/patrimoine-numerique/la-conservation-du-patrimoine-numerique-enjeux-et-tendance.html. [Accessed 14 December 2009]

Cardon, D., 2008. Le Design de la Visibilité. Un Essai de Cartographie Du Web 2.0. *Réseaux Sociaux de l'Internet*, *Réseaux*, 154, Paris: Lavoisier.

Carou, A., 2007. Archiver la Vidéo sur le Web. *BBF*, 2007, 2, p.56-60, Available at: http://bbf.enssib.fr [Accessed 14 December 2009].

Corroy, L. ed. 2008. *Les Jeunes et les Médias. Les Raisons du Succès*, Paris: Vuibert/INA.

Crepel, M., 2008. Les Folksonomies Comme Support Emergent de Navigation Sociale et de Structuration de l'Information sur le Web. *Réseaux Sociaux de l'Internet*, *Réseaux* n. 154, Paris: Lavoisier.

Cosenza, G., 2008. *Semiotica dei Nuovi Media,* Milano: Bompiani.

De Souza, C., 2005 *The Semiotic Engeneering of Human Computer Interaction*, Cambridge, Massachussets: MIT Press.

Eco, U., 1979. *Lector in Fabula,* Milano: Bompiani.

Eco, U., 1984. *Semiotica e Flosofia del Linguaggio*, Torino: Einaudi.

Ertzscheid, O., Gallezot, G., Boutin, E., 2007. *Perspectives Documentaires sur les Moteurs de Recherche: entre Serendipité et Logiques Marchandes.* Available at: http://archivesic.ccsd.cnrs.fr/docs/00/17/21/69/PDF/ertzsgallbout.pdf. [Accessed 18 December 2009].

Georges, F., 2009. Représentations de soi et iIdentité Numérique. *Web 2.0. Réseaux* n.154, Paris, Lavoisier, pp. 165-193.

Gervais, J., 2007. Web 2.0. *Les internautes au pouvoir*, Paris: Dunod.

Getting Beyond Big Online Video, Comscore, 2009

Hayles, K., 2007. Hyper and Deep Attention: the Generational Divide in Cognitive Modes. Available at: http://www.mlajournals.org. [Accessed 18 December 2009].

Intelligence Demographic Charts on You Tube, Hitiwise, Decembre 2008

Jacquinot-Delaunay, G. & Kourti E., 2008. *Des Jeunes et des Médias en Europe*, Paris: L'Harmattan/INA

Le web 2.0 et l'essor des contenus audiovisuels sur Internet. La Lettre du CSA, n° 204, avril 2007, Available at: http:www.csa.fr/actualite/dossiers/dossiers_detail.php?id=122697&chap=2971 [Accessed 10 December 2009].

Les Nouvelles Formes de la consommation des images: TNP, TVID, VoD, sites de partages, piraterie, analyse qualitative, QualiQuanti (Daniel Bô, Claire-Marie Lévêque, Alexandra Marisglia), Paris, CNC, novembre 2007, Available at: www.cnc.fr/Site/Template/T1.aspx?SELECTID=2809&ID=1898&Type=4&Annee=2007&t=2 [Accessed 10 December 2009].

Manovich, L., 2001. *The Language of New Media,* Cambridge Mass.: MIT University Press.

Manovich, L., 2008. The Practice of Everyday(Media) Life, Available at: http://www.manovich.net/DOCS/manovich_social_media.doc [Accessed 10 December 2009].

Online Video. New Face of the Internet, Comscore, 2008

Peirce, C.S., 1978. *Collected Papers*,.Cambridge Mass.: Harvard University Press. Translated from english by M. Bonfantini, 2003.Milano: Bompiani.

Rouquette, S., 2009. *L'Analyse des Sites Internet.* Bruxelles: de Boeck Université, collection Médias-Recherche.

Stiegler, B., Giffard, A., Fauré, C., 2009. *Pour en Finir avec la Mécroissance*, Paris: Flammarion.

Uricchio, W., 2009. Télévision: l'Institutionnalisation de l'Intermédialité. In M. Berton & A. K.Weber, ed., *La Télévision du Téléphonoscope à YouTube. Pour une Archéologie de l'Audiovision*, Lausanne: Editions Antipodes.

Vernet, M., 2007. Les Archives de Cinéma et d'Audiovisuel et les Bibliothèques. In BBF, n° 2, p. 5-11, Available at: http://bbf.enssib.fr/ [Accessed 18 December 2009].

Agnese Vellar

"Lost" (And Found) in Transculturation. The Italian Networked Collectivism of US TV Series and Fansubbing Performances

1 Television Audiences in Networked Publics

The internet has evolved since the late 1990s from text-based technology to the multimedia and visually communication technology of today. During the 1990s, television fans adopted social media such as Usenet newsgroups to communicate with like-minded people, thus participating in the construction of *audience community of practice* (Baym, 2000). In the 2000s the reach of the extended to a much broader public and social media such as web forums, blogs and social network sites became popular among young people. Adopting and adapting social media, young people are now participating in the construction of *networked publics* (Ito, 2008; boyd, 2008) that are both digital social spaces and imagined communities. These technological and social changes are affecting the way television fans consume media products, communicate with like-minded people and participate in the construction of collective identities. In fact, fan cultures are evolving from site-based communities to a *networked collectivism* (Baym, 2007).

To investigate how Italian television audience participate in networked publics I conducted ethnographic research on the networked collectivism that emerged around US TV series, using the case of the community of Italian Subs Addicted (ItaSA). ItaSA is a fan organization that produces amateur subtitles (*fansubs*) for US TV series. ItaSA developed a Web 2.0 portal with a web forum and chat channels where Italian fans can interact. Staff members are young adults who work in teams to produce subs without expecting monetary reward. Their products are consumed by a young audience that is not satisfied by Italian national television which broadcasts a dubbed version of the US TV series long after the US distribution. The aim of this paper is to describe how members of the generation Post (Cohort 1979-1991), by adopting and adapting technologies and media content to fulfil spectatorship needs that are no longer satisfied by national broadcasters, participate in the construction of a generational imagined community in the networked publics.

2 Generations of Fandom: from Cultural Dupes to Networked Amateur Experts

Fans are consumers with an intense engagement with a mass media product. For a long time they have been represented as *cultural dopes* both by mass media and by social researchers. This negative view of fans is a result of the moral dualism which exists between the high-taste and rationality of the intellectual élite and the perceived low-taste of fans (Jenkins, 1992). The academic view of fans as deviants changed during the 1990s when Jenkins proposed conceptualizing fans not as isolated people, but as productive consumers involved in creative and collaborative social practices that are part of a participatory culture. Then Baym (2000) and Hills (2002) conceptualised communicative and productive practices of fans as performance from which emerges communities and cultures. However, the way fans participate in the construction of digital social space and collective identities is evolving with the rapid changing of the technological and media environment. In the evolution of television participatory culture three different generations can be identified: (i) the subcultural fandom of the 1980s, (ii) the online fan groups of the 1990s, and (iii) the networked collectivism of the 2000s.

2.1 Conceptualizing Fan Practices: from Textual Poaching to Performances of Fan Audiencehood

Textual Poachers by Henry Jenkins (1992) is considered to be the book that launched fan studies – an academic field that investigates fan practices from an ethnographic approach and that contributed to the re-conceptualization of media fans as productive consumers. In *Textual Poachers* Jenkins describes a television subculture that emerged around sci-fi TV series. Jenkins defined fans as *textual poachers* because they appropriate professionally produced television text and create derivative works such as *fanfiction* and *fanvids* that expand the narrative universe, often with alternative meanings. Jenkins thus focused on the fans' relations with the media text and conceptualised the amateur production as a form of resistance to the media industry. Studying the online fandom that emerged in Usenet newsgroups in the early 1990s Jenkins re-conceptualised fans as *textual hackers*, leaving behind the oppositional interpretation and focusing more on the pleasure fans experience in decoding complex series such as *Twin Peaks*, which satisfies their need for cognitive control over the text.

Baym (2000) and Hills (2002) conducted deeper investigation of the interpersonal and emotional dimension of online fan practices. Baym conceptualises fan groups as *audience communities of practice* that emerge from informative and interpretive discursive performances. The characteristics and the atmosphere of online groups depend on (a) the technical affordances of the medium that fans use to communicate, (b) the media genre and (c) the characteristics of the participants. Focusing both on the psychological and the socio-cultural dimension, Hills describes online fan practice as self-performance of audience-as-a-text. Online fans are *textual performers* that compete and collaborate to acquire (i) cultural capital (skills, knowledge and distinction), (ii) social capital (a network of friends, acquaintance, and professionals) and (iii) symbolic capital (fame, accumulated prestige and legitimation of other conjunctions of capitals). In performing their fan audiencehood they gain pleasure reliving the cathartic moments previously experienced during the fruition of the original text, and, at the same time, they create a second order of text that other fans can consume.

Fan studies of online groups describe how a techno-élite of educated people with a particular interest in mass media contents adopted mediated technologies as the Usenet newsgroups to discuss, pool perspectives, share knowledge and collaboratively analyze text. The online social interaction between fans can be interpreted as performance of fan audiencehood that a public of lurkers can consume and from which collective identities can emerge.

2.2 Productive Consumers in the Networked Publics: a Collectivism of Amateur Experts

In the 2000s, the internet and television converged in a cross-media platform where corporations distribute transmedia storytelling intended to involve consumers in a narrative universe (Jenkins, 2006a). At the same time youth and young adults of the post industrial countries are adopting a new generation of social media such as Facebook, MySpace, Twitter and YouTube, known as Social Network Sites (SNSs) (boyd & Ellison, 2007). Youth use social media to keep in contact with friends (*hang out*), experiment with new form of learning and self-expression (*mess around*) and connect with like-minded people in interest-driven networks (*geek out*) (Ito, et al., 2009). Young adults use Facebook[1] to keep in contact with old friends and acquaintances and thus crystallise ephemeral relationships and accumulate social capital (Ellison, et al., 2007). Television fans publish self-produced and derivative works in content sharing sites such as YouTube[2] or deviantArt[3], thus competing with professionally produced media texts. The quality and the amount of amateur content raise a debate on the legitimacy of amateur work and, at the same time, on the possibility to capitalise on it. In fact corporations aim to create brand communities and

---

[1] http://www.facebook.com/
[2] http://www.youtube.com/
[3] http://www.deviantart.com/

stimulate amateur production that promotes the brand itself. At the same time mass media present audience productivity as a form of *piracy* because consumers appropriate copyrighted material to create derivative works.

Fan groups are adopting SNSs thus evolving from site-based communities to a *networked collectivism* (Baym, 2007) of *amateur experts* (Baym & Burnett, 2009). A networked collectivism emerges around one or multiple texts and is distributed throughout a variety of offline environment and online sites, which may be amateur blogs, profiles in SNSs or professional web portals. Through creating amateur content and participating in a networked collectivism fans promote their media passion. However, they do not ask for economic reward because they are satisfied by other kinds of reward which are social (relationships that can be of use for future career opportunities) and cultural (discovery of new content). The networked shape of fan groups raises many issues concerning the coordination, the coherence and the efficiency of the productive and communicative practice of fans. However, SNSs give fans many opportunities to share multimedia material and keep in contact with other fans in a global digital environment. In cross-media platforms transcultural flows of amateur and professionally produced content are emerging. Consumers thus have access to popular content produced in different countries and can develop new forms of cultural competences, which have been defined as *pop cosmopolitanism* (Jenkins, 2006b). Contemporary research on television audiences should thus investigate the risks and the opportunities raised by the changed relationship between media industries and productive consumers, by the adoption of SNSs by fans, and by the emergence of a global interactive environment where audiovisual content flows across nations.

3 Italian Audiences, US TV Series and Networked Publics

Different generations of Italian audience grew up watching foreign products as US TV series (Grasso, 2007; Scaglioni, 2006). In fact Italian networks broadcast TV series that US producers sold at low cost to foreign countries (Williams, 1974) because it was cheaper to buy these than produce their own fiction programmes. However, the consumption habits of Italian audiences are continuously evolving. Young people are no longer satisfied with the way Italian networks broadcast TV series, both because the dubbed versions that are broadcast do not give them the opportunity to appreciate the original dialogues and because they are broadcast long after they are shown in the US. This dissatisfaction is particularly true for the members of the generation Post (cohort 1979-1991) who have grown up through their teenage years watching US TV series and integrated new media in their consumption habits. Aroldi and Colombo (2003) point out that older Italian generations have different narrative models and gain pleasure from television consumption by immerging themselves in a fictional world (*fictional model*) or by decoding TV texts (*artificial model*). The Post generation, on the other hand, stigmatises the medium of television and prefers interactive technologies because of their performative aptitude (*simulator model*). Thus they thus have used p2p networks to download digitalised TV episodes uploaded by US fan groups (*crew*) working as *rippers*. In Italy this phenomenon emerged during the mid 2000s as a result of the growth of broadband and the great involvement of young audiences in complex TV series such as *Lost*. In fact *Lost* was intentionally designed with a complex plot and narrative hooks, with the aim of involving internet users in an intellectual challenge and engaging television consumers in continuative and repeated viewings (Askwith, 2007). Italian fans cannot wait for national networks to broadcast dubbed episodes and so they download original episodes as soon as the US crew share it on p2p networks. Since many Italian fans do not understand English well they use amateur subtitles (*fansubs* or *subs*) produced by fan groups such as Subsfactory[4] and ItaSA[5]. These fan groups are hierarchically organised volunteer communities that produce derivative works such as amateur subtitles (*fansubs*) and thus Italianize US fiction and work as mediators for the Italian audience (Barra, 2009). Fansubbers can be defined as *amateur experts* because they are involved in a time-consuming activity and because they are developing professional expertise without earning anything from their work. Italian professionals started to capitalise on the skills of fansubbers to produce professional content without always acknowledging their contribution[6]. In the Italian mediascape the relationship between professional and amateur culture is evolving and should thus be investigated. To better understand the role of the fansubbing community from the perspective of the fans themselves I conducted ethnographic research on the Italian participatory culture that emerges around US TV series. After an explorative research on the Italian networked publics I studied the ItaSA community with the aim of investigating (i) the role of amateur activity in the life of the most enterprising fans and (ii) the emergence of collective identities in Italian networked publics.

---

4 http://www.subsfactory.it/
5 http://www.italiansubs.net/
6 For example MTV Italia co-opted subbers to produce subs for videos shared on the online site QOOB (http://it.qoob.tv/). ItaSA subbers also work as unpaid television presenters for the mobile TV programme SpoilerTV produced by La3Tv. Furthermore, professional adaptators have on occasions informally contacted subbers for advice about how to translate dialogues that contain references to previous episodes that the professionals hadn't seen. Subbers also claim that some professionals use their amateur Italian adaptation to create the dubbed version of episodes which are then brodcast on commercial TV.

3.1 An Ethnographic Study of the Italian "Starring System"

My study focuses on how participatory culture in networked publics can be understood as a *starring system* – a network of individual and collective performances of fan audiencehood. In the starring system fans compete, collaborate and remix cultural material in order to gain visibility and acquire social and cultural capital. In performing fan audiencehood on networked publics fans produce a second order of audiovisual text that can be consumed by an invisible and potentially broad audience. Focusing on the social dimension of fan culture makes it possible to describe how Italian audiences participate in the construction of a contemporary media environment and negotiate their role with the media industry.

To investigate Italian participatory cultures I conducted a 20month ethnographic study (March 2008 - November 2009) of networked publics focusing on the linguistic imagined community of Italian speakers and, in particular, on the community of ItaSA. I combined a multi-sited participant observation and a computer aided qualitative content analysis of multimedia texts. Since I conducted the participant observation on the social context where the members of ItaSA interact, the research field changed with the evolution of the community. In the first few months I interacted mainly in the forum and on the online chat platform. Some of the members then started to use also Facebook and Twitter[7], so I extended my participation to include those SNSs too. My data consists of digital content produced by fans or co-produced with media professionals – online text conversations, hypermedia online profiles, audiovisual fanart, magazine and online articles, radio and TV programmes. Because the quantity of data was so great, I used computer-aided qualitative data analysis software –NVivo – to analyze the material where the members of ItaSA perform their fan audiencehood. I then conducted biographical interviews with 12 of the most participative members of the community (6 male and 6 female) to understand the role of their amateur activity in their personal lives.

3.2 Transnational Fansubbing as Performance of Pop Cosmopolitanism

ItaSA is the biggest Italian fansubbing community – a demanding and time-consuming fan practice that involves a team of subbers in the production of Italian subtitles for US TV series. The team is of three to five subbers and an editor taking the role of coordinator creates the final version of the sub, a textual file

---

7   http://twitter.com/

which is published on the web portal[8]. The ItaSA community was created in December 2005 by Klonny, a 17-year-old boy who discovered Legendaz, a fan-subbing community producing Portuguese subs for Brazilian audiences, whilst surfing the web in search of information about his favourite TV series. Klonny then contacted other fans that he had met on various Italian forums and set up a team of subbers to produce the subs for *Lost, Smallville, Supernatural, Joey, Charmed*, and *C.S.I.* At that time their main aim was to release subs as soon as they could, partly as a challenge to the other Italian fansubbing community, Subsfactory, so since it is easier and therefore faster for Italian *subbers* to translate from Portuguese rather than from English, they worked from the Portuguese versions created by Legandaz.

Motivated by the challenge to produce subs quickly, they then developed their language skills by learning both English and Portuguese and started to collaborate with fansubbing communities in such countries as France and China. Thanks to *word of mouth*, other fans asked to collaborate, partly with the aim of developing their language skills but also for the gratification from the productive practice itself. Subbers enjoy creating Italian dialogues because they feel as if they are closer to their favourite characters and, at the same time, they can express their creativity by personalizing the original dialogues. The small group of fans evolved into a hierarchy of 199[9] who have produced the subtitles for 250 TV series, anime and independent movies (7 Admins, 19 Seniors, 16 Publishers, 80 *Traduttori*[10], 68 *Traduttrici*, and 9 Synchers[11]). Most of the staff are young adults, in particular university students from computer science and communication faculties. Subbers with the role of admins are more technologically oriented and developed a web portal with a graphical homepage and Web 2.0 functionality, in this case a recommendation system and a personalised interface which works like an amateur programme schedule that Italian fans can use to choose content to watch. The portal integrates a wiki with a collaborative television encyclopaedia, a collective blog where fans publish news, spoilers and their own reviews. When the fan group evolved into a larger staff, subbers started to care more about the quality of the subs and to spend time not just on the translation of the original dialogues but also on the adaptation of the textual and cul-

---

8   Subbers only publish textual files containing subtitles on the portal. The portal has no links or any reference to video files that are shared in p2p networks.
9   Quantitative data related to the staff are updated to November 2008, three years after the creation of ItaSA.
10 *Traduttori* is the Italian for *male subber*, while *Traduttrici* is the Italian for *female subber*. It is interesting to note both that a gender difference is made clear and that the roles are defined in Italian. Staff members prefer to use the Italian *Traduttori* and *Traduttrici* rather than the English *subbers*.
11 A syncher creates different versions of the sub to adapt it to different versions of video files that circulate on p2p network.

tural references that the US TV series contains. Since subbers are first and foremost fans, they know the characters and the storyline of the series better that many professional adaptors. However, an episode of a TV series may also contain numerous references to American culture, e.g. the youth culture of *teen dramas* such as *Gossip Girl*, the geek culture of *The Big Bang Theory* or the lesbian culture of *The L Word*). To adapt terms that come from the pop US cultures they make use of online collaborative resources, for example slang dictionaries such as Urban Dictionary[12] or wiki encyclopaedia such as Wikipedia[13] or Lostpedia[14].

Collaborating as amateurs, the staff members have accumulated linguistic capital (English and Portuguese comprehension skills), social capital (that of friendships with fans from different Italian local communities, online relationships with other fansubbing communities and acquaintances with national professionals) and cultural capital (i.e. textual knowledge about plots and characters of specific TV series, intertextual knowledge about specific genres and extratextual knowledge about the culture that is depicted in a series). Subbers' accumulation of capitals has led to a growth in their popularity within Italian networked publics: the ItaSA community has grown to total of 155.392[15] users. Subbers are now recognised as experts by fan cultures and by mass media[16]. They are thus motivated to spend time producing subtitles not only by the symbolical capital they gain but also by the gratification of re-enacting the actors' performances in their favourite series. The productive practice of fansubbing could thus be interpreted as a performance of pop cosmopolitanism that allows fans to relive the cathartic emotion of the viewing and to perform their cultural competences. Since some subbers have become *micro-celebrities* in networked publics they also function as role models thereby stimulating younger fans to get involved in productive activity. In 2009 at least 30 fans a week asked to become subbers. To maintain the quality of the subs the staff thus developed a formal process of selection and tutorship. Fans must first pass a subbing test, after which they become *Traduttori junior*, and then after a trial month of demonstrating their commitment they officially become *Traduttori*. Editors monitor the quality of the work of the members of their team, assessing their ability in

---

12 http://www.urbandictionary.com/
13 http://www.wikipedia.org/
14 http://lostpedia.wikia.com/
15 Quantitative data concerning the community are updated to 1st December 2009, four years after the creation of ItaSA.
16 ItaSA has been described as an amateur community of experts in the mobile tv programme *SpoilerTv* (La3Tv), in webradio shows such as *Versione Beta* and *Dispenser* (Radio2), in national magazines such as *Wired Italia* and in national tv programmes such as *Sugo* (Rai4).

English comprehension, Italian writing and the use of editing softwares such as VisualSubSync which have been adapted to produce subs.

In four years ItaSA evolved from a small group of fans to an online hierarchical organization with a formal education system and an interdisciplinary staff. At the same time the forum evolved into a Web 2.0 web portal. When the community became popular the professionalised subbers struggled to reconcile the amateur ethos that motivated them to participate in a fan community in the first place with the highly structured organization that ItaSA became. Because of that some seniors attempted to involve the new subbers in an online social life so as not to lose human contact during an highly formalised online productive process. Some subbers, however, left the community because it had lost its original amateur ethos. In Klonni's case, four years after he founded ItaSA, he decided to leave because he was beginning to acquire morals more common in a multinational, and then founded Stubborn Italian Jackass[17], a new fansubbing portal. This could be interpreted as a step in the evolution of the Italian fansubbing system from a duopoly to an oligopoly. However, we should remember that the forms of capital exchanged in this marketplace do not produce monetary compensation – they lead instead to social relationships and transcultural flows from which collective identity emerges.

3.3 *Constructing Identity in Networked Publics: ItaSA, Itasiani and their Audiences*

ItaSA is an online platform but at the same time it is a social space where two different collective identities have been constructed: the official and multi-sited identity of ItaSA and the emergent and networked identity of the Itasiani. The official identity of ItaSA has been constructed by the staff in the media, in the web portal and in the official SNSs profiles. In the media they present themselves as a generational audiencehood stimulating an identification process in the young Italian audiencehood[18]. To differentiate themselves both from digital piracy and from the professional cultures they stress the fact that they do not get paid for their work. At the same time they distinguish themselves from *regular* fans because of their high involvement in both analytical and productive activity. Furthermore, they visually present themselves as a professionalised community. In fact the portal was designed with a Web 2.0 aesthetic with the

---

17 http://www.jackassubs.com/
18 In the pilot of SpoilerTv's mobile Tv show, ItaSA introduce the show with these words: «Who among us who grew up during the 80s can say they never watched their favourite tv series like MacGyver or Hazzard in the morning when skipping school?». The staff creates threads in the forum to share memories of their first television passions experienced during the 1990s.

aim of looking like a professionally produced site. To reconcile their amateur ethos and their professional competences they present themselves with *performance of humor* (Baym, 1995). For example they define the time that they spend producing subs as *wasted hours* and when asked why they work free they regularly answer «we've got a screw loose»[19]. ItaSA uses also SNSs to publicise their activity. They created a Facebook group[20] and a Twitter account[21] to keep fans updated with the latest released subs and the latest published blog entries. The ItaSA Facebook group has 10.755 subscribers posting wall to wall expressing their gratitude for the amateur work of the subbers and also complaining when they are not fast enough to release the subs.

In the media, in the official SNSs profiles and on the portal the ItaSA staff perform their official identity in front of an audience of fans that consume their products. However, the staff also developed social spaces inside the portal with the aim of involving users in participatory activities. The portal integrates chat channels and a forum with 1.493.236 posts and 51.010 threads structured in 32 thematic boards. The forum is moderated by the most participative users, who take on the official role of Moderator and coordinated by a Senior (a subber with an organizational role). Informative and interpretive practices are performed in the forum using both Italian and English. In fact users share English information (sometimes translating it into Italian) and comment in Italian on TV series but integrating English words and slang expressions learned while watching TV series[22]. Chat channels have been created by subbers to collaborate in real time for the production of subs. However, users also join them to hang out with other fans. Since 2008 Facebook has been adopted by a broad Italian population and a linguistic imagined community has emerged. ItaSA users adopted Facebook to communicate with fans that they met on the forum. In Facebook fans communicate in Italian while in Twitter they also use the English language since Twitter is still not perceived as a national community. The subbers use their Facebook personal profiles to notify the community when new subs are released, to share organizational information with other members of the team and also to comment on the TV series that they are watching. Subbers also create Skype collective video calls to watch episodes together. Finally subbers organise official offline

---

19 These expressions are English translations of the Italian «ore perse» and «ci manca qualche venerdì». I asked them to translate the sentences for this article since they use them to present themselves.
20 http://www.facebook.com/group.php?gid=140795750240&ref=ts
21 http://twitter.com/italiansubs
22 For example Italian fans integrate the term "XOXO" (hugs and kisses) in their online conversations because they have picked it up from watching the teen drama *Gossip Girl*. It was used by US young people in online and mobile messages and then incorporated into the *Gossip Girl* idiolect.

meetings at least once a year, while people in particular geographical areas (such as Milan or Rome) meet each informally other more often.

From the ongoing interaction between staff members and the users in different online and offline sites there emerged (i) a sense of belonging that characterises the broad collective identity of the Itasiani (as they call themselves), (ii) interpersonal relationships between the staff members that evolved into offline local tight-knit groups and in romantic relationships and (iii) online friendships between users that are maintained through different social media. The Itasiani are thus a collective identity that emerged in an online forum and that is now distributed in multiple SNSs and offline environments and that has a central role in the broader networked collectivism of US TV series.

4 Conclusion: the Pop-Élite and the Transcultural Networked Collectivism

In this paper I've described the emergence of a networked collectivism of TV fans who adapt a cross-media platform and produce flows of derivative contents that are consumed by a national generational audience. In the networked collectivism I've observed two main different forms of participation: (i) the adoption of p2p networks and social media by a generational audience to fulfil their entertainment needs and (ii) the collaboration between individual fans and fan groups who are adapting both digital technologies and professionally produced fictional contents. Italian fans adopt Skype, Twitter, Facebook, chats and forums to communicate with peers and other fans. Interacting in those environments they hybridise Italian with the English language and with cultural references to US TV series thereby constructing a symbolic system that works as a common ground for an Italian generation that doesn't identify itself with the previous television generation that emerged around the Italian broadcasting system. The most enterprising fans spend time selecting, translating and adapting US TV contents. Collaborating online they construct new social structures such as translocal friendships, fan groups and hierarchical organizations such as ItaSA.

The Italian networked collectivism should be interpreted as a transcultural and generational imagined community. This emerges from a process whereby cultural forms such as TV series move through space (*transculturation*), undergo *hybridization* by merging with local features e.g. the Italian language so that the foreign becomes domesticated, or *indigenized* (Lull, 1995). Italian fans of USA TV series domesticate foreign content and digital technologies through the construction of a symbolic and social environment that they perceive as *their own*. This is a generational identity characterised by (a) a networked social structure due to the adoption of SNSs (technological dimension), (b) a pop cosmopolitanism atmosphere as a result of interest in foreign products and in US cultures (media dimension) and (c) a performative aptitude that is typical of the Italian Post generation (characteristic of the participants).

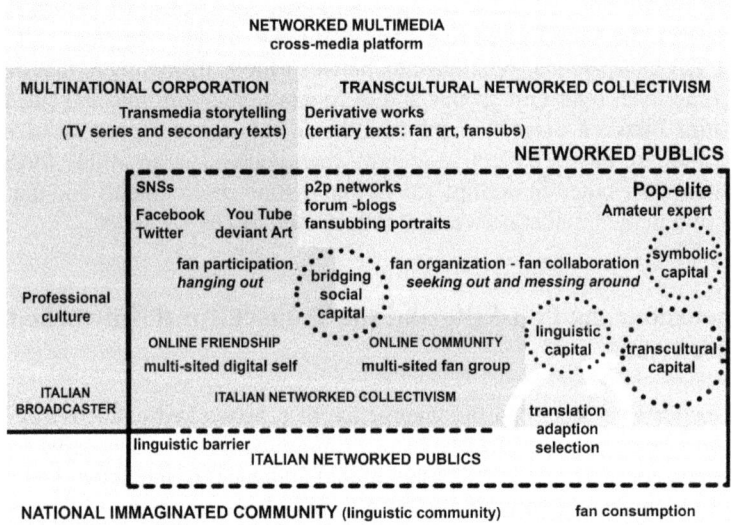

Figure 1: Italian collectivism in the networked multimedia (Source: the author)

Participating in the networked collectivism fans acquire (i) linguistic capital to translate foreign content for the Italian linguistic community, (ii) transcultural capital (knowledge and skills about foreign products and cultures; distinction from preceding Italian generations) which they use to create amateur products that appeal to a pop cosmopolitan generation, (iii) bridging social capital to develop relationships with other Italian and foreign fans and (iv) symbolic capital acquired from popularity on networked publics and, at the same time, legitimization by the mass media. They could thus be defined as a pop-élite of amateur experts who are adapting the networked multimedia to fulfil their personal needs as fans, thus producing an entertaining platform that an Italian generational audience is adopting (Figure 1). At the same time they work as role models for younger fans who are thus stimulated to get involved in collaborative activity.

To further investigate how Italian audiences construct their generational identities in networked publics we should thus compare the forms of participation from a transmedia and transcultural perspective. In fact, while corporations hybridise media forms to create engaging (and branded) storytelling, audiences construct their individual and collective identities by using different languages, participating in multiple sites and domesticating foreign cultures. However, new language skills, new forms of literacy and new socio-cultural competences are required to participate in the digital global environment. In this article I've described the emergence of a pop-élite that has also a formative role for younger fans. We should thus investigate the educational opportunities that are emerging in fan groups, with the aim of understanding how the creation of derivative content could enhance youth media literacy and cultural competences and whether collaborative project around pop culture texts can be applied in traditional educational contexts.

Bibliography

Aroldi, P. & Colombo F. eds., 2003. *Le Età della tv. Indagine su Quattro Generazioni di Spettatori Italiani*. Milano: Vita e Pensiero.

Askwith, I.D., 2007. *Television 2.0: Reconceptualizing TV as an Engagement Medium*. Masters thesis. Boston: MIT.

Barra, L., 2009. The Mediation is the Message: Italian Regionalization of US TV Series as Co-creational Work. *International Journal of Cultural Studies*. 12(5), pp. 509-525.

Baym, N. K., 1995. The Performance of Humor in Computer-Mediated Communication. *Journal of Computer-Mediated Communication*, [Online] 1(2). Available at: http://jcmc.indiana.edu/vol1/issue2/baym.html [Accessed 03 April 2008].

Baym, N. K., 2000. *Tune in, Log on: Soap, Fandom, and Online Community*. Tousand Oaks, CA: Sage.

Baym, N. K., 2007. The new shape of online community: The example of Swedish independent music fandom. *First Monday*, [Online]. 12 (8). Available at:
http://firstmonday.org/htbin/cgiwrap/bin/ojs/index.php/fm/article/view/1978/1853 [Accessed 27 October 2009].

Baym, N. K. & Burnett, R., 2009. Amateur Experts: International Fan Labor in Swedish Independent Music. *International Journal of Cultural Studies*. 12(5), pp. 1-17.

boyd, d. & Ellison, N., 2007. Social Network Sites: Definition, History, and Scholarship. *Journal of Computer-Mediated Communication*, [Online]. 13(1), Art. 11. Available at: http://jcmc.indiana.edu/vol13/issue1/boyd.ellison.html [Accessed 15 October 2008].

boyd, d., 2008. Taken Out of Content. American Teen Sociality in Networked Publics. Ph. D. Berkeley: UC - Berkeley. Available at: http://www.danah.org/papers/TakenOutOfContext.pdf [Accessed 21 December 2008].

Ellison, N. B., Steinfield, C. & Lampe, C., 2007. The benefits of Facebook 'friends': Social capital and college students' use of online social network sites. *Journal of Computer-Mediated Communication*, [Online]. 12(4), Art.1. Available at: http://jcmc.indiana.edu/vol12/issue4/ellison.html [Accessed 04 November 2008].

Grasso, A., 2007. *Buona Maestra. Perché i Telefilms sono Diventati più Importanti dei Cinema e dei Libri*. Milano: Mondadori.

Hills, M., 2002. *Fan Cultures*. London: Routledge.

Ito, M., 2008. Networked Publics: Introduction. In: K. Varnelis, ed. *Networked Publics*. Cambridge, MA: MIT Press. Available at: http://www.itofisher.com/mito/ito.netpublics.pdf [Accessed 15 June 2009]

Ito, M. et al., 2009. *Hanging Out, Messing Around, Geeking Out: Living and Learning with New Media*. Cambridge: MIT Press.

Jenkins, H., 1992. *Textual Poachers: Television Fans & Participatory Culture*. New York: Routledge.

Jenkins, H., 2006a. *Convergence Culture: Where Old and New Media Collide*. New York: New York University Press.

Jenkins, H., 2006b. *Fans, Blogger, and Gamers. Exploring Participatory Cultures*. New York: New York University Press.

J. Lull, 1995. *Media, Communication, Culture: A Global Approach*. Cambridge: Polity Press.

Scaglioni, M., 2006. *Tv di Culto. La Serialità Televisiva Americana e il suo Fandom*. Milano: Vita e Pensiero.

Williams, R., 1974. *Television: Technology and Cultural Form*. London: Wesleyan University Press.

Leopoldina Fortunati

Digital Native Generations and the New Media

1 Introduction

The aim of this chapter is to further our knowledge of digital native generations. I will draw my data from two research projects, one carried out in 2008 to investigate the relationship between social participation and the new media (namely the computer/internet and the mobile phone) and the other carried out in 2010 to investigate the relationship between the second digital native generation and the new media. The data resulting from the first research were unexpected, probably because of the particularly flexible methodological instrument I used - the essay - while the second set of data is the result of research I carried out with the specific purpose of inquiring further into the unexpected results of the first research.

This chapter is organised as follows: in the opening section I will present and discuss a socio-psychological approach to the notion of generation, on the basis of which I will examine the recent notion of the digital native, originally presented by Prensky in 2001. The second section will describe the method used in the research, while the third section will offset out the results of the research. A final discussion will conclude the chapter.

2 Reflections on Generations

What is a generation? In sociological terms, Gallino (1993, pp.318-319) suggests that a generation is an ensemble of individuals who were born in the same space of time measured in a half-decade or decade who are the object of specific social actions and evaluation and are subjected on the whole to social, cultural and psychological experiences that are similar, although bearing strong differences as a result of different class affiliation. As regards the temporal range of a generation in western societies, this can be estimated as about twenty years. Dilthey (1875) proposed a historical notion of generations and considered a generation an ensemble of individuals who may be of different ages but who lived at one and the same time an historical, determinant experience: later examples of this would be the generation of 1968 or the generation of the resistance against fascism. Another notion of generation is the genealogical or parental view which takes as being a generation each level of ancestry and descent of a subject, so a generation is seen as that of the parents, the grandparents, the children and so on. This notion has a high information value especially when the individual of reference suffers a radical change, as in the case of migrants or entrepreneurs coming from the working class or a background of farm working. In these cases, those who belong to the second or third generation exhibit specific behaviours that are different from those of the individual of reference.

The sociologist par excellence on generation is Karl Mannheim, who in his essay *The Problem of Generations* (1952) describes how people who are located in the same generation may see the world in a specific way, very differently from their counterparts from previous generations. Rather than considering new generations as repositories of societal norms, Mannheim underlined the potential dynamism of new generations by portraying them as subjects challenging existing societal norms and values and bringing social change. The idea of each generation sharing a vision of the world is the premise that sometimes explains social effervescence or change, since each generation starts from where the previous generation arrived and goes ahead trying to improve the world which they have inherited. There are three main points to Mannheim's theory of generation, as identified by Turner and Edmunds (2002).

1) Not only classes but generations too can express self-consciousness. A particular cohort of individuals can unify itself into a self-conscious age stratum and give a collective response to a traumatic event or catastrophe such as a war.

2) Not only classes but generations too can be agents of social change and can organise a cultural and political alternative to the status quo.

3) Not only classes but generations too can produce structures of knowledge or consciousness that express particular location.

This sense of belonging to a generation occurs even though a number of differentiated generation units who fight one another can be found within each generation and even though conflicting views of reality can exist within the same generation, in part due to the diversifying power of other variables such as gender, ethnicity, social class and so on. Research on life-course and ageing would serve to lend insight to the question of generations.

Each new generation not only feels more the spirit of the age, the Zeitgeist, as Kortti in this same volume stresses, but is particularly sensitive to social change. Moscovi (1976) argues that young people have the future in their hands and for this reason they are generally more disposed to change the world than older adults and the elderly. Through the development of consensual universes of meaning - he continues - each new generation is able to re-shape a collective awareness by explaining objects and events in such a way that makes them accessible and thus responding to the immediate interests and needs of the generation members. Of course identification with a generation is not automatic and depends on the individual. There must be the will on the part of an individual to adhere to a generation and there are individuals who are keen to disown them. Obviously the identification with one's own generation is easier when one is young and when the generation in question is powerful and capable of capturing the spirit of the age. Difficulties come into play as time goes by and identification with a specific generation is no longer seen as attractive. The persistence of generational identities can be fragile, as the research on social movements show when it describes what happens to young political militants when they reach middle age (Scott & Marshall, 2005).

The issue of identification with one's own generation is also related to identification with one's own age. On this point Jung's work is very enlightening. In his essay *Anima and Death* (1970, p.155) Jung argues that disowning one's own age is akin to rebelling against one's own death. Both represent the non-will to live, since the non-will to live and the non-will to die are the same thing. Generally one remains attached to one's own past in the illusion of remaining young. Growing old is becoming extremely unpopular: people in contemporary societies have difficulty in understanding that it makes no sense to refuse to grow older as it makes no sense not to grow out of infancy. While people express pity towards a person in his thirty who is still infantile, they consider a youthful seventy-year-old to be 'cool'. However, states Jung, both are perverse and psychologically deformed. Today people frequently seem younger or older than their real age. It is as if consciousness has slipped from its base and no one is able to keep pace with natural time. The same thing happens with identification with one's own generation, especially when this generation gets older.

All the conceptualizations of generations contribute to outline, as we will see below, different but important dimensions of digital native generations which are the specific object of our study. In the last century several scholars expressed an increasing interest in investigating the contribution of generations to social change. An author often neglected in generational studies is Sigmund Freud, who however gave a very important contribution to the question of generations. Freud (1970, p.213) in the last essay of Totem and Taboo, *The Return of Totemism in Childhood,* argues that there is a collective soul where mental processes develop as they do in the individual soul. And it is this collective soul that guarantees continuity in the psychic life of humankind, since it allows us to ignore the breaks of psychic acts caused by the death of the single individual. Without presupposing this notion of collective soul there would be no collective psychology, or psychology of populations. If the psychic processes were not transmitted from one generation to another, but each generation were obliged to build ex novo its attitude towards life, no progress or evolution would be possible.

But how important is the psychic continuity in the sequence of generations? And then which instruments does a generation use to convey psychic conditions to the following generation? These two questions are still open since the two first answers which come to mind – direct transmission and tradition – are far from being adequate explanations. Continuity is in part assured by the heredity of psychic dispositions, but, to become effective, they need to be stimulated by personal and social events in individual lives.

The transmission of psychic dispositions to succeeding generations has to negotiate with a principle which challenges it - that is, a new generation starts, as we mentioned before, from where the old generation arrived and then develops a new stage. In his book *Children of the Great Depression,* Elder (1974) demonstrates exactly this and shows how children brought up in an age of great austerity have a completely different view of the world from those raised in a situation of economic prosperity (showing attachment to work, being disposed to personal sacrifices, exhibiting a sense of the duty and so on). This is one of the core problems which jeopardise the economic future of industrialised countries. The generational replacement by new generations in the organization of labor in these countries has struggled against such problems as the devaluation of material labor and the consequent need of immigrant labor, and the lack of ability or interest shown by children of entrepreneurs to replace their parents in the management of factories and so on.

Another question still open is the understanding of to what extent the socialization of successive generations differs and to what extent values and behavior are discontinuous between generations. Harding (1933, pp.206-219), however, proposes an original model in this regard. She argues that although human problems remain the same through different ages, their form and content change. A rhythmic movement develops, more spiral shaped than cyclic, in which two moments can be distinguished: one, long and slow, which corresponds to the evolution of human consciousness from primitive times to the modern age; and the other, relatively more rapid, which corresponds to the change from decade to decade. On the former spiral the change which takes place in the space of a generation is infinitesimal; on the latter spiral the change is more consistent and enough to lead elderly people to say that they do not understand young people's behavior any more or how the world works. But sometimes a cultural form changes suddenly and with revolutionary violence. When this happens, the values of parents and grandparents are turned upside down and a true fracture develops between generations, for example, the generational revolution that has taken place in moral and sexual behavior since the first world war. But young people do not understand that they cannot resolve or destroy within themselves the attitudes of their ancestors through a mere change of their conscious or intellectual point of view. While new ideas and ideals take possession of the conscious sphere, attitudes which are refused fall into the unconscious where they exercise an influence which no one can escape.

A last very important concept elaborated by Harding as a principle of a genealogical conceptualization of generations is that parents might give children a certain independence, but on closer inspection young people must actually win their independence. If it is parents who give it to children, then young people are 'in debt' and cannot be independent. They must win the right to be themselves and become adults. Stealing the right to become adult is a "necessary crime". As Harding states, in myths the acquisition of individuality and personal autonomy is often represented by the theft of something very precious which was belonging to the reserve of the gods. This is the case of Adam and Eve or the case of Prometheus who steals divine fire and brings it to earth.

A more recent approach that could help a lot in this analysis of generations is looking at generations as a social construction carried out by multiple social forces (Aroldi & Colombo, 2003). However, it is inside the intimate relationship between parents and children with all its conflicts and its necessary mediations that the social construction of what is an important part of a new generation is carried out. Generally a new generation is built on the refusal of parental authority and on the identification with one's own peers on the one hand, and, at the same time, on parents' capacity to gradually distance themselves from their children while trying to empower them. Durand (1960), looking at literary movements, observes that historically one movement is succeeded by another opposite one. As a reaction to parents' paths, tastes and authority the new generation will build a completely different movement. These two factors – refusal of parental authority and identification with peers – lead the new generation to create an alternative social space, language and their own values and culture so that their life is particularly different to that of their parents. The sharing of all these elements on the part of subjects who were born approximately at one and the same time constitutes the cultural and social core of a generation. It strengthens the sense of belonging to the same generation even more. The mechanisms that push a sizeable number of people to identify with a generation are manifold. To cite a few mechanisms: the sharing of a common social memory, the same cultural products, the same strategies carried out in the educational process and the same political and social events and environment. At the same time a generation has to deal with a society which has become very different to that experienced by the previous generation because of effervescent social change. So, the concept of generation is one of those that capture better this sense of social change and movement.

3 Digital Native Generations

This socio-psychological path has enlightened how the concept of generation is multidimensional-temporal, historical and genealogical - and refers to a social actor who through a sense of belonging and identification with one's own generation is capable of producing social change and dynamism and of developing consensual universes of meanings. Although each generation is more able than the previous one to feel the spirit of the age, the identification with one's own generation as well with one's own age is not exempt from contradictions and ambiguity. Speaking of generations also means – as Freud argues – dealing with the transmission of psychic activity from one generation to a new one and with the use of instruments which are activated in this transmission. This transmission is partly unconscious and partly conscious, and it may be subject to the conflicts existing between different generations who may have opposite interests and visions of the world. Harding's model addresses the problem of the extent to which socialization, value and behaviours between generations are different and it is very useful to capture the rhythms of the social changes from generation to generation. Finally, the notion of generation is analyzed as a social construction. In this framework how can the notion of digital native be accommodated?

Prensky introduced the term digital natives in 2001 with the purpose of describing the generation of young people who were born in the late 80s and grew up in environments saturated by all kind of digital technologies. These are young people so accustomed to the new media that they can be considered "natives" of the digital world, while their parents, who did not encounter the digital media until adulthood are considered "digital immigrants". Middle-aged and elderly people, who in the past were the guides of technological development, are becoming extraneous to electronic technology and to the information science of the digital world. Although they have been using communication and information technologies since they encountered them at a certain point of their life, they find it difficult to accept and to appropriate them and are thus called digital migrants. The digital divide between so-called digital natives and digital migrants poses questions regarding important aspects of the social life: the transformation of family life and of the relationships between parents and children, models of socialization and the effectiveness of educational strategies at primary and secondary level and so on. Scholars and journalists have created many terms to express the characteristics of these young people. Some examples of these labels might be: "Nintendo Generation" (Green, Reid & Bigum, 1998); "Net generation" (Tapscott, 1998; Oblinger & Oblinger, 2005); "The Millenials" (Howe & Strauss, 2000); "Digital natives" (Prensky, 2001); "Generation Me" (Twenge, 2006). Common topics in these conceptualizations are, for example, the tech-savviness of this digital generation in dealing with and using digital technology, a self-perceived uniqueness, a strong multitasking capacity and an increasingly global vision.

The definition of digital native has been criticized by many for being too generic and misleading (Bennett et al., 2008). Of course, it is misleading because the concept of generation is multidimensional, as we mentioned above, while the notion of "digital native" is mono-dimensional and suffers from technological determinism. Furthermore, the appropriation of digital media, which has changed so many daily practices and routines of the population of industrialised countries, has spread in a differentiated way not only among countries, but also inside individual countries among different social strata. However, I have kept the term digital native generation as a label to be explained by these research projects, with the hope that it will be given a more plausible meaning. Of course it would require a number of research projects to outline the main sociological, political and economic characteristics of this digital generation and to give substance to this analysis. This study, in which the use of the new media is considered as a cultural form, might represent a point of departure. In effect it is just the extremely rapid change of this cultural form that has made the theme of the digital native generation emerge dramatically because it has produced a sociocultural fracture between generations.

Furthermore, in this research the notion of digital generation mirrors quite well the convenience sample involved since it was made up of students from the undergraduate Multimedia Sciences and Technologies programme and of the graduate Multimedia Communication programme at the University of Udine (Italy), who are very much present online as prosumers (Toffler, 1980).

4 Method

The materials analyzed here come from two research projects that I carried out in Italy with the general purpose of understanding how digital native generations relate to the new media. In particular, the aim of the first research project was specifically that of investigating the relationship between social participation and the new media. But this collection of essays (N=156) has produced unexpected results on the issue of generations. Each year I put together a collection of essays around the main topic of communications and this was the first time that in a collection young students expressed their concerns regarding the relationship that their younger brothers and sisters had with computers, the internet and mobile phones.

Two years later I organised a second research project in order to investigate this question in particular. Sixty-six students were involved in this second project. The large majority of this group was made up of undergraduates born in the late 80s, i.e. young people who can be considered "digital natives". I asked them to write an essay on the relationship between children/teens and the new media, asking them to use first names only in order to guarantee anonymity. On the whole I collected a corpus of 222 essays. This corpus has reached an adequate degree of saturation in the sense that after a certain number of essays no new data and only already acquired information emerged. This method is directed at the centre the subjective dimension of respondents and aims to identify the meaning of certain discourses. It is mainly descriptive as it collects data as personal documents which are analyzed in all their richness and individuality. The criteria that I applied during this research in order to ensure a degree of reliability are those discussed by Bryman and Burgess (1994): apart saturation, the other criteria are reactivity, internal acceptance, which regards the position of the researcher in the relationship with the respondents, internal coherence, which concerns the final product, which should exhibit and justify the choices made and the path followed during the research and, finally, completeness, which regards the synthesis which should reflect the complexity of the research undertaken.

Essays like those collected here might be seen as the answer to an open question. The choice of using this tool was dictated by the need to avoid as much as possible the so-called effect of social desiderability, typical of the answers in questionnaires and interviews (Corbetta, 1999), and to have an instrument which would allow more expressivity and spontaneity in the content. The texts obtained contain very little 'reactive' information so they are influenced to a very limited degree by the interaction between the researcher and the object of study and are less influenced by distorting effects. The texts collected were broken down into categories of discourse with the purpose of capturing the most relevant discursive frames on the relationship between the digital native generation and the new media, and were studied by means of content analysis (Altheide, 1996). I calculated the frequencies of the categories singled out to see the most recurrent ones, but I also retained and discussed, as Silverman (1993) suggests, categories with relatively low frequency but significant for enlightening or clarifying some points of the analysis. This allowed me to trace a conceptual map of the discourse outlined by these students on the theme investigated, to be used to analyse and describe the most important issues discussed. This type of content analysis is a fairly non-intrusive and very flexible methodology (McNeill & Chapman, 2005).

Although these kinds of texts might be analysed from many aspects, here I limited my analysis to some of the most relevant issues outlined by students on the characteristics of digital native generations and on the main features of their relationship with the new media.

5 Results

The main result of these two research projects is that the group of students involved in the first research project identified a second digital generation that is emerging. And in effect if one sees a generation as lasting two decades it is appropriate to talk of a second digital native generation. This second generation, made up of the younger siblings of the first digital native generation, can be identified by a series of characteristics that also mark the first generation. Giovanni says:

> "I see in fact how the younger generations are more at ease with the use of the new media and, although they do not have useful information knowledge, they are able to manoeuvre with ability in the meanders of the net. As the new generations were born immersed in the new technologies and so, like a child born in that water, they are predisposed to swim in it, while those like me who have seen the internet, the interfaces and the various Windows progressively spreading should however be facilitated by the fact that they have both basic information and the historical memory to analyse and live the media in a more complete way."

The new digital generation is more at ease with the new media but they do not have the historical memory of the evolution of the new media which the first generation has. The historical experience of changes undergone by technological artefacts is important as it gives more awareness of the path of innovation. What are the main features of this second digital generation in its relationship with the new media? A theme which recurs frequently in these essays is the immoderate use or abuse of the new media on the part of teens. Michele writes:

> "Teens make excessive use of the videogames and are lost in front of the computer. These disadvantages hinder their mental development but especially the physical development. Being a basket coach I have observed that children are increasingly awkward and they are not able anymore to do even the simplest movements…For them all these technologies are indispensable, they are not able anymore to do practical things and for them it is very difficult to learn. Furthermore, the majority of teens smoke or drink to feel like adults or to not being excluded from the group."

Lorenzo adds in this respect:

> "the majority of teens use exceedingly this extraordinary means of communication and do not use it in a useful and constructive way; on the contrary they use it exclusively for amusement and fun…."

Francesco :

"the enormous expenditure of time that young people today dedicate to the use of the new technologies. The mobile phone, for example, is set aside only while they sleep".

This abuse of media has precise social consequences as Sara writes:

"A problem connected to the use of the new media is the possibility that their use during the daytime of children and adolescents prevails to the detriment of the time which they spend in concrete spaces of socialization and meeting-places, such as sport or leisure centres".

Another negative consequence is that children and teens lose mastery of body movements and gestures as they are too sedentary: Marco explains:

"I live in a very crowded neighbourhood where once young people used to meet each afternoon to play football. I remember that we were able even to stop the traffic, if an action had to be concluded. Such a scenario today is very rare if not completely absent. One cannot see children playing on the street as the majority of them prefers to spend their afternoon in front of the computer (Internet or videogames). This, I think, it is a very negative factor as playing a real football match is different from a virtual one. In a real football match players have fun, fight, want to win at any cost. A situation like this helps a guy growing up and I think it is positive as it allows the child to move and to relate really with the others."

Ambra too shares Marco's opinion:

"Many times, a child who is not a great football player, renounces to play, as he prefers to overcome another level of difficulty in this car videogame instead of engaging, working hard and maybe losing".

This abuse of the new media by the second digital generation is perhaps also connected to the fact that this generation is more precocious than the previous one. Niccolò writes that his little sister texts all day with her friends and has had a broadband connection since she was just 16 years old. And Giacomo states:

"I have a cousin who is 14 years old and I see that her approach towards these technologies has been quicker in comparison to mine."

Serena writes that:

"often, surfing on the Internet one finds teens between 10 and 15 who subscribe various social networks… and who pose as twenty-years-old, publishing sexy pictures with the purpose to seem more mature and obtain more visits to their profile".

Michela reports:

> "my sister who now is 7 years old is able to use well the computer... Many times I have to take her away from the PC or the TV and I try to make her doing things that do not create addiction from the technologies..."

Alessandra writes that while she subscribed to a social network for the first time when she was 18, her brother did so at 12. Matteo is especially struck by the fact that 8 or 9-year-old children have a mobile phone and he writes:

> "children will develop increasingly less the oral communication also with their parents, because they send hundreds of texts per day".

There are certainly some positive aspects to this precocity inasmuch as it means that an accelerated learning speed means that any service can be accessed incredibly quickly, but there is also the negative aspect of children and teens acting as if there were adults which brings with it potentially dangerous consequences. Francesco writes:

> The four-year old daughter of one of my acquaintances, having seen only once how the application which reproduces animals cries for the I-phone is downloaded and works, is able to do it. The current new generation has a high capacity of learning, connected more to a direct use rather than a theoretical knowledge. This precocity can be seen as a gift but it might bring negative aspects. The risk is to find us surrounded by adolescents who behave as adults.

Giulia writes about her 13-year-old brother who began to use a computer at age10 and is very quick in learning and using the new media, for example the mobile phone, and several times he has explained to his sister some difficult things about technologies and software. Cristiana thinks that this technology used so early is:

> "detrimental for the development of human relationships as a virtual social relationship cannot be compared with the relationship developed in contact with the others".

The general thinking expressed by these respondents is that the new generation approaches the Internet and the mobile phone too soon. Even before the age of 13, many have free access to the Internet and to a mobile with the consequence that they also have a precocious encounter with sexuality/pornography (Livingstone & Bober, 2005) or with inappropriate content in general. Matteo writes:

> "This morning it happened that I heard some teens who were talking about sexuality in a disarmingly normal way, citing friends who watched porno films or sex scenes in websites."

Sara states that the new media:

> "present a complex reality in which find a place behaviour and content which might be harmful for the development of the youngest children."

Another theme which identifies a shared opinion is that this new digital generation is increasingly isolated from true real/human relationships in favour of digital contacts and it is undergoing a strong process of de-socialisation. Even when there is not an abuse of the Internet, the simple daily use of it leads to isolation and closure. Morris, looking at his sister, states:

> "...she seems to increasingly 'closed into herself' while she looks forwards to 'open' the computer to enter in relationship with the others, the problem that I notice is that she prefers talking to others through the computer rather than face-to-face."

Of the same opinion is Andrea, who writes that if the use of these media is not moderated they will increasingly damage interpersonal relationships. Vittorio, too, observes that children and teens are losing the capacity to communicate 'body-to-body':

> "they pass a long time in front of social networks... oral communication with them becomes increasingly difficult as they are used to be always right and they cannot put up with listening to other people outside the PC...this is discouraging because the reality in which they live is as if it was the PC and the Internet."

Vittorio's observation is interesting since he stresses an important limit of computer-mediated communication. If one encounters somebody who argues against you online, any need for negotiation and mediation can be avoided with a simple click. The a-synchronicity of computer-mediated communication along with the almost absolute control of one's own communication process contributes to create generations incapable of mediating and negotiating with others. Daniele says that:

> "teens meet on forums as once teens met around the fountain in the evening, near a little wall, in the churchyard or in the football field. They didn't have a specific purpose, they were pushed by the only pleasure to stay together. Can we compare these two different ways to stay together? In my opinion we cannot, but it is exactly this that the sedentary and the spread passivity of young people are creating... Teens publish their pictures, often half-naked and use these virtual squares to approach the other sex."

Angelo adds that many adolescents and teens:

> "use these media with the purpose to take refuge in themselves, avoiding the direct contact with real people, building purely virtual relationships and not learning anymore to face the glance of somebody their age or the shiver of an agitated discussion in which participants raise their voice. The new media are an easy way out for those young people who are very

shy or feel to be inadequate towards the society surrounding them and they will probably have difficulties when life will force them to undertake socialisation activities such as, for example, a simple job interview."

Marina states:

"I guess that the relationship that the second digital generation has with the new technologies is dramatically limiting concrete social relationships. The virtual interpersonal relationship does not coincide with real experience, as our senses are not involved in the relationship with the other (who is not present)."

Sara N. even argues:

"I have a 15-year old sister and I see that at her age either one has Messenger or My Space or Facebook or you are excluded from the group".

Sara's consideration tells us that for this generation being online in some social networks becomes the premise for acceptance by the body-to-body group.

Another important issue that emerges in several essays is the solitude of this second generation in front of the screen. Sara writes:

"They are alone in this digital world without anyone to keep an eye on their decisions, behaviour, choices on the net…"

Parents are too busy at work and consequently they are not able to take proper care of their children who grow up quickly and without a point of reference, an adult role model to follow. Or because, as Stefano states:

"the diffidence and the ignorance of the current adults towards these media do not contribute to helping young people learn how to deal with the Internet to the Internet to transmit to young people."

In some cases parents intervene actively in the education to the new media, for example, by limiting the time available for the use of the computer by their children, as in the case described by Fabio who has a 14-year-old brother. But in general a total lack of media education is described in these essays. Marco states that parents control too little and leave too much freedom to teens in media use. Gianmaria adds:

"parents should stimulate the critical capacity of their children before letting them surf on their own in Internet alone, since this is the only way to prevent their children from being influenced by thousand of stupid and non-educational advice".Lisa is convinced that the internet is a new powerful means of education. Cristiana writes that all the teens she knows subscribe to a social network, "which assumes the role of the 'competent adult' who helps them to go through adolescence".

On a similar line, Irene says:

> The Internet imposes itself on children as educator, because, since it is the resource which has a hold on their daily life with greatest frequency, it delivers information regarding everything you might need...The role of educator becomes confused with that of the new technologies. As in everything, balance is the right compromise. It is absolutely necessary that proper education from parents merges and completes with the "instructions for use" delivered by the Internet. Furthermore, the child whose parents are not familiar with the Internet uses it in total anarchy...Maybe when the future educators-parents (even if we grant for the sake of argument that they would not abdicate this role ... given the bad examples of current parents ...) are able to handle the Internet with sufficient mastery, they will guide their children towards a well-balanced use of the Internet.

However not all these students are pessimist regarding the relationship between the second digital generation and the new media. Many appreciate the positive characteristics and what the Internet and mobile phones can offer. Filippo proposes this metaphor to describe the new digital generation: those who were born and live in the world of buttons instead of the senses. Alessandra writes that despite what is often said, the Internet might be a great resource for young people who know its languages and what it can offer, although

> "also among them information literacy is often lacking. They use the Internet in a selective and limited manner..."

On the whole, the observations of the first digital generation are the symptom of the fact that they perceive the fragility of their parents in guiding their young children in the relationship with the computer and the internet. This fragility derives from the fact that the parents, being digital migrants, do not have the appropriate knowledge and experience to help their young children in establishing a reasonable relationship with the computer and especially with the internet.

There is a frequent request of protection towards the new digital generation. Gianmaria writes that:

> "it is necessary to protect children and adolescents from a content which might be inappropriate as regards the psycho-physical development as well as from violent, illegal or harmful content such as online pornography and paedophilia-pornography".

But there is also a very frequent request of media education or education to technology, as Mirko claims specifically. It is as if these groups of students perceived a profound empty space. Federico states that:

> "it is to be hoped that the new generations are educated from the beginning to the correct use of these new instruments, with the purpose not only to be able to use them in their complexity but also to minimise the problems of harmfulness that can derive from them."

In the same way, Giulia writes:

> "there is a need of a concrete work on the part of the main educational agencies, that is the school and the family, and of the other institutions with the purpose to protect new generations from abusive contents".

These concerns are also the symptom of a double important change in the conceptualization of the role of young generations. These young students of the first generation of real digital natives frequently state that they assume a kind of parental role towards their younger siblings. Concerns relate to the care of younger children are in fact part of the behaviour and engagement of the parental role. The novelty in the attitude of concern on the part of this group of students is that, surprisingly, they do not exhibit any intolerance of taking a caring role in the relationship with the younger siblings, which has changed from previous generations (Wellman & Wortley, 1989). Much research shows how young people in previous generations did not appreciate parents forcing them to take on the responsibility of caring for their younger brothers and sisters, a role that they perceived to be more properly that of their parents (Heywood, 2001).

At the same time as the first and the second digital native generations take on a role of caring for younger siblings, they also take on a parental role to their own parents. The recent caring attitude of this generation of young digital natives has developed spontaneously, probably because they have perceived that their parents, as members of one of the last analogical generations, do not know enough about computers and the internet and do not know how to deal with education of the new media. In the space of a generation grandparents have lost the control over technologies. Fabio writes that:

> "grandparents ask grandchildren to tune in or fix the TV, to adjust the thermostat and set the digital clock. On the other hand, children learn directly from parents, who put them in the front of television when they are few months old, show them family pictures on the compute and give them a mobile phone while they are at primary school."

Nicola also writes that in many cases it is the new generations that give an 'inverse' education to the adult group on the use of the new media. Alberto underlines how family roles have changed: on technologies the roles have inverted and it is now children who teach their parents. Fabio, Nicola and Alberto describe how now it is parents who ask for help when they need to make a computer or other devices work.

6 Final Discussion and Remarks

This study is based on two research projects involving convenience samples of students, so these results are not at all generalizable and must be taken with great cautiousness. Having said this, for the restricted situation investigated the research shows that a second digital native generation is emerging which is made up of the younger brothers and sisters of the first digital native generation. This second generation seems to stress the characteristics of the first generation. For the first time in my classes undergraduate and graduate students have expressed a concern about the relationship that their younger siblings have with the new media (internet and mobile phones). These young people do not believe that their young brothers and sisters have the appropriate cultural tools to successfully handle all the opportunities and the dangers that the internet presents. Furthermore, they observe that their young brothers and sisters may spend too much time in front of the screen, may become sedentary, are too precocious, may become isolated from social relationships in co-presence and thus lose their skills of socialization. They think that their siblings are left alone to explore the complex world of the new media without being given any proper education regarding the new media. Since many children and teens are exploring the virtual world alone, having been self-taught rather than have been given any 'digital' guidance or education, they are asking for protection and guidance from their older brothers and sisters from the first generation of digital natives.

Secondly, it is clear from the students who are first generation digital natives that a cultural fracture, to take up the term used by Harding, has occurred between this generation and the previous generations of digital immigrants, namely their parents and grandparents: adults and elderly people are no longer able to teach young people: what happens is the opposite. This digital generation has to take on a kind of parental role towards their younger brothers and sisters and also fall into a parental role to their parents and grandparents. This phenomenon seems to have a certain strength given that it has also been observed in other countries such as Belgium, where Claire Lobet-Maris collected similar evidence[1]. Furthermore, among the various dimensions that make up the notion of generation this research shows how the diffusion and use of the new media have created changes in the genealogical dimension in particular, involving the relationships between parents and children, grandparents and grandchildren, and brothers and sisters of different ages.

Thirdly, the identification of the first and the second digital native generations is not so strong because rather than sharing specific, historical events, they share devices, technologies and their uses and practices (Bourdieu, 1990), as well as strategies of domestication in their everyday life (Silverstone & Hirsch, 1992;

---

1 Personal communication.

Fortunati, 2009). While social or political events generally produce emotional shocks which strengthen the subsequent process of identification with the related generation, the practices of the use of technological artifacts create similar languages, competences, complicity, but identification remains more superficial, assuming there are no significant social or political events producing a strong effect on a generation.

Fourthly, from this research it emerges that parents give a lot to these digital native generations: technologies, freedom, sexuality. But as Harding explains very well this is a bad starting point inasmuch as these generations have to take on the role of adults in order to win these riches. The social construction of these generations seems to be made by parents with scarce authority and presence and by children destined to remain weak because they are not pushed to exercise the will to "steal" the fruit of adultness.

Finally, this analysis of digital native generations should only be seen as a starting point for further broader research which looks at sociological, political and economical aspects and which studies cross-generational problems that the diffusion and appropriation of digital media open up in the light of systematic comparisons with other countries.

Bibliography

Altheide, D.L., 1996. *Qualitative Media Analysis*. Thousand Oaks: Sage.

Aroldi, P. & Colombo, F. eds., 2003. *Le Età Della TV. Indagine su Quattro Generazioni di Spettatori Italiani*. Milano: Vita & Pensiero.

Bennett, S., Maton, K. & Kervin, L., 2008. The 'Digital Natives' Debate: A Critical Review of the Evidence. *British Journal of Educational Technology*, 395, pp.775-786. Available at: http://www.cheeps.com/karlmaton/pdf/bjet.pdf [Accessed 20 December 2009].

Bourdieu, P., 1990. Structures, Habitus, Practices. In P. Bourdieu, *The Logic of Practice*. Stanford, CA: Stanford University Press.

Bryman, A. & Burgess, R.G., 1994. *Analyzing Qualitative Data*. London: Routledge.

Corbetta, P., 1999. *Metodologie e Tecniche della Ricerca Sociale*. Bologna: Il Mulino.

Durand, G., 1960. *Les Structures anthropologiques de l'imaginaire*. Paris: P.U.F.

Edmunds, J. & Turner, B.S., 2002. *Generations, Culture and Society*. Buckingham: Open University Press.

Elder, G.H. Jr., 1974. *Children of the Great Depression: Social change in life experience*. Chicago: University of Chicago Press.

Fortunati, L., 2009. Theories Without Heart. In: A. Esposito & R. Vich eds. 2009. *Cross-Modal Analysis of Speech, Gestures, Gaze and Facial Expressions*. Berlin: Springer, pp.5-17.

Freud, S., 1913. Totem und Tabu. Über einige Übereinstimmungen im Seenleben der Wilden und der Neurotiker. Translated from German by C. Balducci, C. Galassi & D. Agozzino, 1970. *Totem e Tabù altri saggi di antropologia*. Roma: Newton Compton, pp. 61-208.

Gallino, L., 1993. *Dizionario di sociologia*. Torino: Tea, UTET.

Harding, E., 1933. *The Way of all Women*. New York: Longmans.

Heywood, C., 1991. *A History of Childhood*. Cambridge: Polity Press.

Jung, C. G., 1933. Anima e Morte. In: C.G. Jung, 1970. *Realtà dell'Anima*. Torino: Boringhieri.

Kortti, J., 2010. The Problem of Generation and Media History. In: F., Colombo & L. Leopoldina, *Media and (Broadband) Society*. Berlin: Peter Lang.

Livingstone, S. & Bober, M., 2005. *UK Children Go Online: final report of key project findings*. 6. London, UK: London School of Economics and Political Science.

Mannheim, K., 1952. The Problem of Generations. In: Mannheim, Karl (edited by Paul Kecskemeti): *Essays on the Sociology of Knowledge*. London: Routledge and Kegan Paul, pp. 276-320.

McNeill, P. & Chapman, S., 2005. *Research Methods*. Philadelphia, Taylor & Francis, Inc.

Moscovici, S., 1976. *Social Influence and Social Change*. London, Academic Press

Prensky, M., 2001. Digital Natives, Digital Immigrants. *On the Horizon* MCB University Press, 95. Available from: http://pirate.shu.edu/~deyrupma/digital%20immigrants,%20part%20I.pdf [Accessed 20 December 2009]

Scott, J. & Marshall, G., 2005. *Dictionary of Sociology*. Oxford, Oxford University Press.

Silverman, D., ed. 1997. *Doing Qualitative Research. Theory, Method, and Practice*. London: Sage.

Silverstone, R. & Hirsch, E., 1992. *Consuming Technologies: Media and Information in Domestic Space*. London: Routledge.

Toffler, A., 1980. *The Third Wave*. London: Collins.

Wellman, B. & Wortley, S., 1989. Brothers' Keepers: Situating Kinship Relations in Broader Networks of Social Support. *Sociological Perspectives*, 323, pp.273-306.

Vesna Dolničar, Sonja Müller, Marco Santi

Designing Technologies for Older People: a User-Driven Research Approach for the SOPRANO Project

1 Introduction

Contemporary societies are facing two striking trends: widespread population ageing and rapid diffusion of new technologies. Since old age very often carries with it diminished abilities and ailing health, we embrace the potentials of technological advances to enhance health, abilities and social integration. Services based on information and communication technologies (ICTs) are already part of everyday life in Europe. However, the complexity and novelty of many new devices and services threaten many older people with exclusion from their use. At the same time, research has shown that a large segment of the growing number of older people in Europe could be offered services which would radically improve their quality of life, but only if services and the ICT components supporting them were designed in such a way as to make appropriate use of ambient intelligence and the new ability of software systems to communicate with users in something approaching natural human to human interaction.

The purpose of the paper is to present a user-oriented research approach and the results of the needs assessment and requirements analysis of the SOPRANO project (http://www.soprano-ip.org/). Service Oriented PRogrammable smArt enviroNments for Older Europeans (SOPRANO) is a European Commission 6th Framework Programme integrated project, which started in January 2007 and will go on until October 2010. SOPRANO is a consortium of commercial companies, service providers and research institutes with over 20 partners from Canada, Germany, Greece, Ireland, the Netherlands, Slovenia, Spain and the UK, The project's main aim is to design and develop highly innovative, context-aware, smart services with natural and comfortable interfaces for frail and disabled older people at affordable prices, thereby meeting the requirements of users, their families and their care providers. Supportive environments are based on the concept of ambient assisted living (AAL), using pervasive technologies such as sensors, actuators and ambient intelligence to provide additional safety and security, which will support independent living and social participation and thus improve quality of life. One new AAL architecture supports pro-active assistance based on situational analysis fed by user input and local monitoring.

Thus, when developing a next generation of smart homes based on ambient intelligence, a positive mindset of the project will be followed: the resulting system will not only act in problematic cases, e.g. in the case of a fall or a burglary, and cases of emergency, e.g. health problems or fire, but will focus to the same level on improving the quality of everyday life for elderly people. As such, the SOPRANO system should act as an informed, friendly agent that takes orders, gives advice or reminders and is ready to offer help and get help when needed. Three strands of research and development are to be integrated: (a) stand-alone assistive technology (products designed to compensate for motor, sensory and cognitive difficulties frequently experienced by older adults); (b) smart home technology (networking of ICT in the home environment, with the integration of appliances and devices to provide control of the entire living space); (c) telecare services (applications addressing care-related needs prevalent among older people, with ICT utilised to enable support from professionals and informal carers).

Services utilizing SOPRANO capabilities are thus designed for older people, including those with age-related cognitive changes. To ensure that services fully meet user needs, developers have been working with potential users of SOPRANO systems throughout the project lifecycle, from user requirements analysis, through iterative prototyping, validation of concepts and functionality and usability tests to large-scale field trials involving users in their own homes. For this purpose, the state-of-the-art Experience & Application Research (E&AR) methods were adapted and extend, providing innovative tools to create a new, consistently user-centred design methodology. Namely, when developing technologies to support a special target group it is always necessary to get a deep understanding of the intended user group. What are the needs, intentions, attitudes and knowledge of the users? What about the context of use? In SOPRANO, understanding and being able to model the mental processes and the behaviour of older people is not only important for developing new products and systems. It is also important for using appropriate methods for analysis, design and evaluation. Methods that involve users directly were adapted to be used with elderly people.

Against this background – after outlining the need for ambient assisted living – this chapter first gives an overview of the characteristics and circumstances of older adults that are perceived to have potential influence on the acceptability of ICTs in general and also assistive technologies (ATs) specifically. The chapter then presents user-centred research design (investigating requirements of older users and exploring the design ideas together with the potential end-users) that is based on adaptation and development of the E&AR methods. As a result of the research, key challenges to independent living and corresponding needs older people have in relation to a better quality of life are also presented. The paper concludes by discussing the benefits of the user-driven approach.

2 Contextual Background: The Need for Ambient Assisted Living

Demographic changes and the need for old people to live at home as independently as possible for as long as possible are pushing politicians and experts to develop new telecare services and update existing telecare services for older and disabled people because since the important aspect of ICTs in this context is to provide and support homecare and independent living. Consequences of demographic changes are also raising questions about financial sustainability of countries to ensure and provide sufficient scope of effective and in the long-run cheaper long-term health and social services. As mentioned by Müller et al. (2008), over the last decade, considerable RTD efforts have been pursued by the European Commission, national governments and relevant industries to provide an adequate technology response to the challenges of an ageing society. In terms of technology uses, the so called "independent living" or "assisted living" domain today comprises a heterogeneous field of applications ranging from quite simple devices such as intelligent medication dispensers, fall sensors or bed sensors to complex systems such as networked homes and interactive services. Some are relatively mature and some are still under development.

Emerging ICTs, such as pervasive computing, ubiquitous computing and ambient intelligence have considerable potential for enhancing the lives of many older people throughout the world (Sixsmith and Sixsmith, 2008). The Ambient Assisted Living Joint Programme of the European Union (AAL Joint Programme, 2009 in Sixsmith et al., 2009), set up in June 2008, has an overall objective of using ICTs to improve the quality of life of older people and specifically to:

- allow people to "age-in-place" by increasing their autonomy, self-confidence and mobility;
- support health and functional capability;
- promote active and healthy lifestyles;
- enhance security, prevent social isolation and maintain the support network of the individual;
- increase the resource efficiency and effectiveness of health and social services.

Recent research suggests that a large segment of the growing number of older people in Europe can be offered ICT-enabled support services which considerably improve their quality of life, provided usability of ICT systems can be equally improved. Thus assisted technology can prolong independent and qualitative living at home in the known local community of older people (and the disabled) when providing needed support, services, security, entertainment and so on.

In sum, the societal trends that SOPRANO is responding to are:

- the increase in the proportion of older citizens in the population due to demographic change;
- the scale and type of needs of older citizens which society must plan to meet;
- the rejection of current ICT-based services by many older citizens and the steady deterioration of non-ICT-based service provision in the information society;
- the paucity of available ICT-based services usable by older citizens;
- the difficulty of designing ICT-based services to be usable by older citizens.

3 Acceptability of Information and Communication Technologies and Assistive Technologies

When it comes to more complex systems in particular, the potential ICT might hold in relation to independent living, (e.g. the impact on quality of life for care recipients, will very likely not be great enough for the guarantee sustainable success of ICT-enabled social and medical support services. Experiences from previous research suggest for instance that organisational, cultural and other non-technological issues come into play if ICTs are to be successfully introduced into daily practice (see for example Müller et al., 2008; Oswald et al., 2007). However, also in relation to the technology itself, up to now all too often simplistic assumptions have been made in relation to the needs and aspirations of those whose independence is ultimately to be supported in one way or another, and also of those who provide assistance to them. There is for instance some evidence to show that many older people, despite being in need of help, are wary of giving outsiders intimate insight into and access to their homes (e.g. SOPRANO, 2007a, 2007b). It seems very likely that they would accept technology based help more readily if they had more say in what information was sent out, to whom and under what circumstances. They are willing to accept assistance but they are not willing to give up their independence to outsiders. Also, the complexity and novelty of many systems and devices that have been developed during recent years seem to carry the risk of excluding older people from using these technologies. Thus the social care approach is challenged by a lack of acceptance on the part of potential service users. Despite their promise of increased safety and comfort, service providers find it difficult to persuade end-users to accept technology-based monitoring systems. As a result, appropriate systems are introduced too late or not at all.

Even though the focus of this section is on the acceptability of ATs, we also refer to other computer technologies and technological devices in general). Namely, social aspects and implications of the acceptance of ICTs on the one hand and ATs specifically on the other often overlap in terms of relevance and attitudes towards them.

A number of studies identified the role of psychological barriers to older adults' learning about ICTs and using them, and these include fear and anxiety, lack of confidence, lack of motivation, and negative attitudes. Together with other additional psychological factors, they are discussed in the following subchapters.

3.1 Attitudes to ICTs

Attitudes to ICTs (with attitude being defined as social psychological concepts which refer to one's beliefs, feelings, and behaviours toward an attitude object) have been identified as significant factors in older people's relationships with ICTs. A positive attitude toward ICTs might lead to an attempt to use an ICT, though it may have little influence on one's actual ability to do so. Older adults are more conscientious and are therefore more likely to be concerned with "doing things right" (Cuttler & Graf, 2004); they are also more likely to be afraid of failing, and thus reluctant to try new ways of doing things. However, even though older people tend to have less favourable attitudes towards computers than younger people, most findings show that the majority of older adults have positive attitudes toward computers. For example, even though the SeniorWatch (2002) study showed that 40% of older people (50+ age group) feel quite uncertain around technology there seems to be a strong interest among the oldest segment of the population in learning about new ICTs; namely more than 50% of respondents agree with the statement that they are keen on learning about technological advances and developments.

Studies indicate that the most important aspect in learning new ICT skills is motivation and a desire to learn. Nonusers have less positive feelings about technology than users and are less likely to understand how it might be useful to them.

3.2 Lack of Interest and Motivation Regarding Use of ICTs

While it is true that older adults are slower to adopt many new technologies, and typically require more training to use them, studies in general reflect a willingness of older adults to adopt new technologies. Even though elderly people tend to use ICTs less than other age groups, few studies indicate that older individuals are less interested and more resistant to learning how to use ICTs than are young adults. On the contrary, older adults have been shown to be enthusiastic about learning how to use computers, and especially the internet, when training opportunities are made available (Morell et al., 2004).

Older individuals may think computers and ATs are irrelevant to their daily lives and offer no advantages. Older people are generally more likely to see no personal benefit in accessing new ICT services, and they are more likely to be unaware that potentially beneficial services and interesting/useful content is available. They will have little motivation to use ICTs and ATs unless the benefits of acquiring such skills are readily apparent. Of course, ATs are not universally attractive to, or universally needed by, older adults; some older people (sometimes labelled as voluntary non-users) are aware of the potential advantages, but do not perceive these as beneficial to them.

3.3 Lack of Confidence and Fear of ICTs

Person-centred barriers are often expressed as intense emotions experienced in initial interactions with computer technology. ICT-related fear and anxiety are significant inhibitors for many older people (Richardson, 2006). Lack of knowledge is often seen as an inhibitor to older people's use of ICTs. Cautiousness and disbelief in one's own capability may be an obstacle in accepting technological advances. Higher levels of computer experience are associated with lower levels of computer anxiety. Thus, lack of confidence is one of the reasons why older people have difficulty in mastering new ICTs.

Women, in particular, tend to identify fear of the machine as contributing to a lack of confidence in engaging with ICTs. Fear can be understood in the context of women's continued discomfort with technology which is at the centre of the social construct of gender and technology. Women also tend to construct themselves as 'technophobic'.

A lack of emotional and practical support from others, unfriendly tutors and inability to use computer interface devices such as a keyboard, or difficulty in understanding instructions were all identified as significant barriers to feeling confident in the use of the technology. Computer anxiety generally diminishes across training sessions.

Older people often feel that their ability to learn to use ICTs is hampered by technological jargon, which they consider to be a 'foreign language', designed for engineers, confusing and intimidating.

Within home environments, a strategy for increasing acceptance of ubiquitous computing may be to introduce it gradually to older adults who are still high functioning (e.g. those in their 50s and 60s) and to increase the supports provided by the technologies, at the same time, as motor control and perceptual and cognitive abilities decline (e.g. as they reach their 70s and 80s).

When developing AT, it is important to understand that the majority of older people prefer not to use ATs which are designed especially for older or disabled people. The older population will show a strong tendency to reject new products which have apparently been designed to cater to some kind of 'special need', as many of them do not want to be considered as distinct from the rest of the population. It would therefore be better to promote a simplified device without it being branded as targeting vulnerable or disabled people. Products such as interactive television services are more likely to attract older users if they are offered as a user-friendly basic service with a wide choice of 'optional extras'.

3.4 Willingness to Learn and to Acquire ICT Related Skills

Older people seem to be slightly more reluctant to use technology and, in particular, new technology. Willingness to acquire ICT-related skills seems to be strongly influenced by perceived needs, perceived usefulness and a sense of personal identity (Morell et al., 2004; Rogers et al., 2004).

On the one hand older adults seem to fear the unknown and, on the other they believe there is a lack of information about the functionality of ICTs and how to handle and work with them. The elderly may exhibit a decline in abilities shown to be important for learning and skill acquisition. Proper instructional design that capitalises on intact abilities and compensates for declining abilities holds much promise for proficient novice levels of AT usage.

It is clear that older people aged between 60 and 74 (the younger old) and the older-old (aged 75 and above) can readily acquire computer skills, navigate web sites and maintain these skills over time. This includes various commonly available software packages, and learning memory training techniques via CD-ROMs, to using specialised interactive software for health promotion. However, studies indicate that computer training with older adults takes somewhat longer than with younger individuals because older adults make more mistakes than younger adults when learning to use computers (Melenhorst et al., 2006).

Learning styles may be placed on a continuum that extends from discovery learning (learning by doing) to reception learning (learning by seeing). Learning styles are known to vary across contexts and tasks, as well as across individual characteristics such as age. In novel situations, older adults tend to adopt a reception style of learning whereas young adults tend to opt for a discovery style. On the one hand informal computer–learning with peers or family members increases acceptance and creates an atmosphere of trust and understanding. On the other hand the complicated emotional situation of both parties can lead to conflicts. In short, often a professional pedagogical approach might be more appropriate.

3.5 Loss of Privacy?

The deployment of intelligent monitoring systems has to be sensitive to the individual's wishes and preferences. While the emphasis is on the quality of life of frail older people, the use of technologies that monitor the activities and behaviours of people in their own homes also raises serious ethical issues, such as loss of privacy and information usage. People generally feel that the use of cameras within their homes for the purpose of monitoring for falls or other accidents is intrusive and violates privacy (Richardson, 2006). They prefer the use of cameras where shadows or movements are depicted but one cannot identify the features of the individuals.

However, the elderly have an overall positive attitude towards devices and sensors that can be installed in their homes and generally towards the concept of 'smart homes' if they believe that there might be numerous ways of improving their everyday lives. It is also important that AT users know what type of information is being collected and when, who has access to such information and whether information can be subsequently linked to the user (Rogers et al., 2004).

3.6 Recognition of ICT Quality and its Usefulness

With regard to technologies for the ageing population, the match between technologies on offer and actual user needs is far from optimal. Lack of acceptance and lack of usability and usefulness are often diagnosed as reasons for limited technology diffusion in this area. For example, for people with reduced physical capabilities and increasing sensory impairment, adequate technological appliances grow in importance. Nevertheless, despite attempts by developers and manufacturers of technical devices and systems to make products as user friendly as possible, all too frequently technological devices or particular features thereof are not adequate for the needs and abilities of older people (Comyn et al., 2006; Melenhorst et al., 2006).

An important factor affecting acceptance among older people is the trust individuals have in automated devices. If technology is developed to support the independence of older adults in their homes, it must be reliable. Equally important is the fact that people must be willing to rely on the technology. User decisions to rely on automated devices depend in part on self-confidence in their ability to control the system. Trust is mainly dependent on current and prior levels of system performance and incidence of faults.

Another important factor influencing the acceptance of technology systems is usability. User acceptance refers to the desire that the user may demonstrate in wanting to use the system again over a long period of time and feeling comfortable when doing so because it is functionally satisfactory. User acceptance includes elements of social and practical acceptability. The concept of usability typically focuses on five inherent characteristics that contribute to the usable nature of a device or product. These include learnability, efficiency for the user, being able to remember how to carry out the sequence of events, avoidance of errors (or alternatively the presence of errors that can be corrected easily) and satisfaction.

3.7 Expectations, Needs and Interests of Older People

Technology has the potential to make life easier, to act as a support to communication with family and friends, to assist with health care and to enable individuals to remain safe and functionally independent in their own homes (Melenhorst et al., 2006; Demiris et al., 2004). Relevance has been identified as one of the critical issues in the uptake of a new technology. Literature review indicates that the most common expectations and needs older people might pursue from assistive systems are the following: ATs should support independence and memory, reduce general emotional burden due to anxiety, stress and worry, prevent unwanted incidents or accidents, support memory, prevent situations that create unwanted negative feelings between patient and carer, create comfort and pleasure, improve security and enhance personal communications. Some of these needs are further discussed below.

- Inter-personal communication and reducing isolation: One of the greatest risks in ageing is not necessarily poor health but isolation. Communication with friends, relatives, health care providers and others is crucial to healthy ageing. Advances in ICTs make it possible and affordable for older adults to remain connected to the world around them. Moreover, a new generation of interactive and easy-to-use applications can be developed for caregivers to ensure that their loved ones are safe and well. Health and well-being also include such activities and enjoyment as learn something new, enjoying new experiences, having fun and managing essential personal services such as transportation and meal delivery. Today's information systems enable access to these and other activities.
- Recreation and leisure: Studies show that more than one fourth of older people experience boredom. In this regard, new technologies can offer an enormous range of possibilities by allowing remote participation in leisure and entertainment activities, for example chat-rooms, online games,

participation in virtual communities. In addition, virtual education can offer new opportunities older people, especially in the case of individuals who live in remote places or have mobility problems.
- Improving security: Security seems to be one of the main preoccupations of older people, particularly for those who live by themselves. These people are concerned not only with anti-intrusion alarms, but also with security related to domestic accidents e.g. fire, water and gas leaks. AT also allows older people to programme a set of functions for a day-to-day routine (for example, programming certain tasks before leaving their homes, such as turning off lights, locking doors, closing windows and so on).
- Solving mobility problems: To the older person mobility problems present an obstacle not only in the home but outside as well. These problems could partially be solved by online shopping, online banking services and the implementation of domotic solutions that help automating some day-to-day activities within the homes of the elderly. Examples of this might be controlling the air conditioning and lighting and raising and lowering blinds.
- Healthcare: Most older people are well aware of their fragility. The assurance of knowing that in an emergency there will be somebody who will react gives older people confidence and peace of mind. However, the wish to have someone available at any moment is not only related to physical health but also has a bearing on any psychological problems deriving from a feeling of solitude. Finally, there are other necessities like taking medicine, carrying out simple health checks such as testing heart rate, pulse rate and temperature or other tests related to a particular disease e.g. sugar analysis, measurement of arterial tension r, ECGs etc. which can be done at home by using alarms and biomedical sensors and devices.

4 User-Centred Research Design

The needs of older people must be borne in mind when designing home technologies if they are to be successful in acting as a platform for independence. An extensive programme of user-related research was implemented to give deeper insight into these needs. Users' perceptions were studied in four stages: eliciting user requirements, generating and evaluating design ideas together with older people, prototype component testing and tests/evaluations in real environments. "This is illustrated by Figure 1.

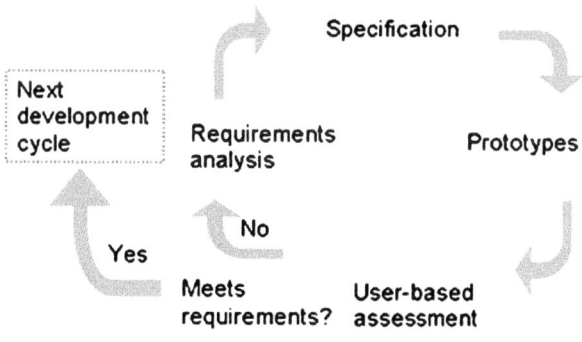

Figure 1: User-centred R&D design (Source: SOPRANO, 2007b)

The approach adopted throughout this cycle was guided by the ecological model and Experience and Application Research (E&AR), which is outlined in the following sections. The adaptation and application of concrete techniques, methods and tools that are appropriate for the involvement of older adults in the research and development process are also presented below.

4.1 Ecological Model for Guiding User Research

While many models of ageing and acceptability of ATs have been developed (e.g. McCreadie & Tinker, 2005; Richardson et al., 2005; Vimarlund & Olve, 2005), an 'ecological' model (Oswald et al., 2007; Sixsmith & Sixsmith, 2000) was deployed as a useful framework for guiding the user research in this project. The ecological approach was particularly suited to the SOPRANO project because it focused on practical aspects of a person's everyday activities, highlighting opportunities for technology and design solutions to support these activities. The model (see Figure 2) draws on the work of researchers such as Kitwood (1993), Lawton (1990) and Sixsmith & Sixsmith (2000). The underlying argument of the model is that the activities that comprise a person's everyday life are shaped by a range of different factors, including attributes of the person and attributes of the immediate and wider socio-cultural contexts. These personal and situational factors operate together in a functional, 'ecological' relationship to facilitate or constrain a person's activities.

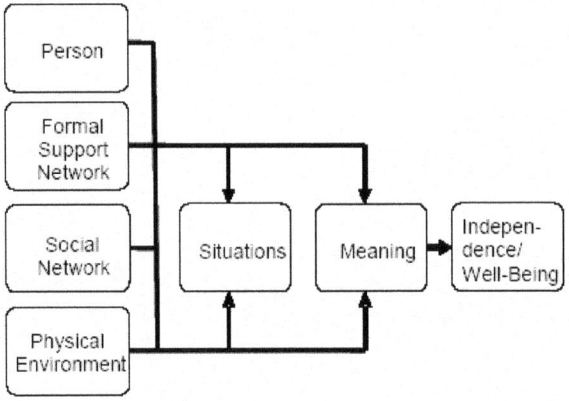

Figure 2: Ecological model (Source: Sixsmith & Sixsmith, 2000)

The ecological approach is useful in looking at the independence and quality of life of older people, because it highlights the impact and experience of age-related dependency within its context (e.g. dependency resulting from cognitive impairment) and allows us to explore how this affects everyday life and well-being. How a person derives meaning from their everyday activities and environment is central to their well-being. Positive well-being results from these factors work together, while – conversely – apparently minor obstacles in any one of them may prevent a positive outcome (Müller et al., 2008). Technological and design intervention can potentially play an important role by ameliorating some of the personal and contextual problems faced by a person who is experiencing cognitive decline. The argument here is that a person's well-being will be enhanced if the intervention facilitates activities that are meaningful and valued by the person and take into account contextual factors of a person's life.

4.2 Experience and Application Research: a Participatory Method of Generating Innovation

Participatory methods enable us to thoroughly focus on users, when they are uses to define user requirements, to iteratively generate design solutions and to evaluate those design solutions in real life settings. Addressing such a target group as older people requires suitable methods. Standard tests might be too tedious, standard questionnaires might not be easily understood by all subjects or it may be too challenging for new ideas to be understood properly (SOPRANO, 2007c). In addition to this, in ambient intelligent systems the context of use of a system is much more relevant than in standard office environments. The project thus needs to develop methods and tools in order to offer prototype context with users. The objective of SOPRANO is not simply to obtain a quick insight into system usage and user acceptance but also to gain an understanding of learning effects which are especially important for applications that are used on a daily basis.

- In the SOPRANO project we applied the concept of E&AR, as developed in 2004 by the Information Society Technologies Advisory Group (European Commission, 2004). E&AR involves research, development and design by, with and for users. It also involves research into methods and tools that will enable this. Both established and especially developed methods based on the latest E&AR approaches are deployed as "innovation enabler", ensuring that SOPRANO systems fully utilise the potential of technology innovation to meet user needs and gain wide acceptance. The possibilities and constraints of ambient intelligence are dependent on evaluations by users in the context of their everyday lives. E&AR must also consider relevance to social and cultural practices. This approach is crucial to the development of human-centred ambient intelligent systems. One of the aims of involving users in ambient intelligence R&D processes is thus to create useful and successful products.
- Working in multidisciplinary teams is fundamental to E&AR. Researchers and developers increasingly recognise the need to cross the barriers of disciplines to create products that match the future demands of users. A more multidisciplinary approach to the development process opens the way to new possibilities, perspectives and methods. SOPRANO has managed to establish a wide net of different disciplines cooperating to pursue its objectives.
- SOPRANO follows the E&AR approach not only by involving users from different European countries throughout the whole R&D process but also by developing new methods and tools that are not only adapted to the target group of the SOPRANO project but also go beyond state-of-the art approaches by developing new methods and tools.

5 Stepwise Approach Towards Elicitation of User Requirements

One of the major objectives of SOPRANO's first user iteration was to involve users in identifying and developing ideas with a high potential for technological innovation. These ideas should be closely tied to the experiences, mindsets and mental models of the users so it was necessary to find a way to elicit user needs. What was important was that the identification of user needs and user requirements as well as the generation of design ideas should not be biased through explicit assumptions or knowledge of what might be possible with today's technology.

- Work started with an extensive literature review. Here the focus was on gaining a better understanding of physical, cognitive and other changes that tend to come with the process of human ageing together with an understanding of the potential impacts of these changes on a person's experience of life in old age. This preparatory exercise was to lay the ground for the identification of key challenges to independent living, with a view to identifying options for the provision of an adequate technological response to these.
- Further to this, a repository of generic situations potentially threatening older peoples' independence or quality of life was compiled on the basis of the knowledge gained from the literature - particularly related to daily activities involving tools and equipment (see e.g. Katz, 1983) - and on the basis of feedback from those project partners with experience in providing support services to older people. This research resulted in some 80 situations that were filtered and categorised. The categorization of these situations revealed three main domains and several sub-domains, with the main categories being: the psycho-social domain, i.e. interaction, communication, recreation, exercising, creative activities, use of communication devices and asking for help; activities instrumental to daily living, i.e. affording independence, reassurance of performing, eating and drinking, meal preparation and cleanup, mobility, sphincter control management, personal cleansing and grooming, shopping, checking body temperature, care of pets, financial management; medical domain (healthcare): preventive measures, monitoring, aid in breathing, sleeping, care of personal devices such as hearing aids, glasses, prosthetics, adaptive equipment, etc.
- The next step in requirements elicitation focused on involving potential users of the SOPRANO system in order to gather their feedback on a) key challenges to independence/quality of life and b) initial ideas on how technology could be harnessed to better cope with these challenges. To this end a qualitative methodological approach was adopted, involving both focus groups and individual interviews. The "situations inventory" compiled earlier in the project was used for triggering responses and stimulating lively discussions among participants. Focus groups as well as individual interviews were conducted by applying semi-structured interview methods. Individual interviews complemented focus groups and were expected to be more effective for some topics and some groups of people (e.g. people with minor cognitive impairments of hearing impairments). The intention was also to compare outcomes of the two methods, i.e. focus groups and individual interviews, in order to contribute to E&AR research.

Overall, 14 dedicated focus groups (involving more than 90 end users) were conducted with different stakeholders (older people, informal carers and care professionals) in the UK, the Netherlands, Spain and Germany. Individual interviews with older people took place in Germany, Spain and the Netherlands. Interviews were carried out in informal and non-threatening settings and were subject to ethical guidelines and procedures for each country.

The work outlined above with older people, informal carers and care professionals revealed a set of key challenges to the independence and quality of life of older people where some kind of technology-based response was seen as having great potential to provide practical help. However, it became equally clear that a 'systemic' perspective needed to be adopted in relation that part of the overall project dealing with technology development. To see the design problem as that of designing a technical system to offer new services without reference to the social world within which the person lives would be to miss a great opportunity – and would very likely lead to failure. A key question was therefore how the technical systems being designed were embedded in the process of delivering new ICT-based services. Based on these considerations, the technical components under development would explicitly be conceived and designed as part of a socio-technical overall system for delivering new ICT-based services to older people. Against this background, a set of use cases and scenarios were developed to effectively capture functional requirements of the technical SOPRANO system by describing interactions between one or more users and the system itself.

5.1 Results: Needs Older People Have to Ensure Better Quality of Life and Development of Scenarios of Use

Key themes that emerged from these activities for further consideration in the SOPRANO design process included both technical and non-technical aspects. Among others, these include the following issues (Müller et al., 2008; SOPRANO, 2007b):

- Social isolation/loneliness: Perhaps one of the biggest problems mentioned by users concerns social isolation resulting in many negative outcomes such as loneliness, depression or the feeling of being cut-off. Another aspect mentioned in this context concerns the feeling of boredom which in many cases seems to be related with a feeling of being socially excluded.
- Safety and security: Another challenge that was frequently mentioned concerns the desire for safety and security. Important issues that were highlighted

in this context include falls, disorientation, control of household equipment or receiving help in the case of emergency.
- Forgetfulness: Forgetfulness seems to be a challenge to independence for many and concerns taking medication, for example, or finding objects in the house. Issues mentioned on the subject of taking medication are multi facetted and relate to forgetting to take the correct medicine at the right time, finding the medicine in the house and also to the unwanted side effects of taking various kinds of medicine. Some people also seem to have problems in managing their appointments or a normal calendar.
- Keeping healthy and active: Challenges were also reported in relation to keeping healthy and active in later life, e.g. when it comes to physical and mental activity and exercise, good nutrition, good routines (such as regular sleep patterns) and, again, taking appropriate medications at the right time. Some people reported difficulties in adhering to specific regimes that have been determined by health professionals, including rehabilitation programmes.
- Community participation/contribution to local communities: During a UK focus group there was strong emphasis on the desire to participate in local government activities and informal and semi-formal support networks. While this view may not appear to be widespread, it points to the view that community participation and contribution to local communities should not be underestimated.
- Accessing information/keeping up to date: Keeping up to date seems to be a crucial issue for many as well. Here, access to local news or the possibility of reading newspapers were emphasised, as was the need to find tradesmen to do little jobs around the home, such as decorating, cleaning and repairs, etc.
- Getting access to shops and services: People often seem to have difficulty getting out of the house for shopping, banking and so on. These kinds of services and support are usually outside the remit of local authorities so that people in need of support have to get such things done on their own, or are dependent on others or voluntary organisations.
- Checking up on care provision: Local authorities may be purchasers and/or direct providers of care at home. As care is provided in the community within a person's own home, it may not always be easy to monitor that the right amount or right quality of care is actually delivered.
- Mobility inside and outside the home: Keeping mobile inside and outside the home was mentioned as a problem area as well. On the one hand this concerns being able to walk as long as possible and on the other hand being able to use public transport. Mobility restrictions were reported to be especially common in winter and in the evenings.

In each case, delivery of improved services supported by SOPRANO systems was seen as having great potential to increase the quality of life of older people and to support their independence, enabling them to stay in their own home as long as they wish. Ideas for improved services were developed in response to each challenge to independence, taking into account not only technological features seen as desirable by focus group participants but also their rejection of other features of technology and of action by service providers seen as intrusive or unnecessary. Criteria to be taken into account in assessment of service ideas were not only the degree of importance to older people but also the level of intrusion e.g. the incidence of false alarms, the load on service providers and the technical feasibility and cost of service delivery.

Scenarios of SOPRANO use were developed based on ideas for new ICT-based services to address key challenges to independence. The scenarios include use cases, which are seen as an effective way of capturing functional requirements of a system via describing interactions between a user and the system. The scenarios not only reflect functionalities of the technical system under design but also the processes, actions and interaction between 'components' of the overall socio-technical system – i.e. informal and formal carers, service providers, doctors, hospitals and so on – and the assisted person himself or herself. The set of SOPRANO scenarios developed within the framework of this first phase of user involvement were:

- "Medication reminding (I and II)" – with the aim of improving the situation of a person forgetting to take medicine.
- "Open door" – with the aim of enhancing safety and security at home.
- "Safe" – with the aim of monitoring activity for signs of problems.
- "Fall" – with the aim of adjusting care to take account of increasing frailty.
- "Easy to use home automation" – with the aim of demonstrating smart home components that support independent living.
- "Exercise" – with the aim of assisting older people in recovery from a stay in
- hospital.
- "Active" – with the aim of monitoring signs of problems and providing support for good routines.
- "Remembering"- with the aim of providing assistance for coping with cognitive ageing.
- "In Touch" – with the aim of combating social isolation.
- "Entertained" – with the aim of countering boredom.

Sixsmith et al. (2009) briefly illustrated one of the scenarios by a simplified and shortened version of a "Remembering" example (see also SOPRANO, 2007b): the older person wishes to leave the house and opens the door to leave. This is recognised by a door sensor. The SOPRANO system reascts by displaying a warning on the GUI (e.g. on a touchscreen near the door) that various appliances have been left switched on in the house (e.g. a cooke or heater, etc) and that the window is open. The GUI displays options for the user (e.g switch off applicances). The user responds via the GUI and/or voice command, either to switch off appliances or to take no action.

The scenarios of SOPRANO use are intended to be entirely specific in time, place, characteristics of the persons involved, their surroundings and in the sequence of events. All participants have names rather than just functional descriptions ("Roger" rather than "the assisted person") so as to enable concise use in textual descriptions and to contribute to a flavor of reality. The reality and plausibility based on specificity were to facilitate critical appraisal by the team and by users. The descriptions have enabled doubts to be expressed, i.e. answers, occasionally negative, to questions such as "Would this be acceptable?", "Would this really help?", "Would this be usable?", "Would this be paid for?" This allowed ideas to be rejected or adjustments to be made in service characteristics. The approach quickly exposed many cases of demand side and supply side weaknesses and fuzziness in thinking producing similar impact. The methodological assumption being made here is that if one specific use case can be found which can be positively assessed, then it is very likely to be an example of many which would also be positively assessed.

6 Methods for Creating and Evaluating Design Ideas Together with End-Users: Multimedia Demonstrators and Theatre Groups

The scenarios represented a key input to the development of a first prototype system, which then underwent an iterative process of refinement involving user feedback at a number of subsequent development cycles. With the results shown above, the first steps within the user-centred design process as implied by DIN EN ISO 13407 are taken, i.e. understanding and specifying the context of use and collecting and analysing users' needs and requirements. The state of the art approach for involving users throughout the whole R&D process is typically followed by the development of prototypes by experts. By doing this, decisions about the conceptual design, i.e. what kind of functions are to be developed and what the interaction should be like, are made by experts. The prototypes, based on these conceptual design ideas, are then evaluated by users. The initial design ideas are, therefore, not based on the mindsets, experiences and mental models of the users but on the decisions of experts. The user can intervene only through the fourth step –user-based assessment.

The E&AR approach applied in SOPRANO changes this situation. Instead of having experts to lead the development of prototypes, a more user-centred design process involves users in the development of the conceptual design and, later, the prototypes. The idea is that it is not users who should respond to ideas of experts but experts who should listen to input from older people and respond to this input. In other words, the main objective of this second stage of user involvement was to validate and further explore use case ideas, utilising multimedia mock-ups and theatre presentations to help users visualise technologies prior to prototyping. Feedback from users was used to refine use cases and finalise requirements for prototypes. The second stage of research thus provided detailed design refinements in relation to each of the 11 scenarios.

SOPRANO used two approaches for prototype testing, both variants of Scenario Based Design (Rosson & Carrol, 2002; Burmester & Machate, 2003):

- design idea generation methods: eliciting design ideas from end users.
- design idea evaluation methods: involving users in refining design decisions taken in previous activities.

One of the main challenges was that potential users would have to create design ideas and evaluate a system not yet in existence. The project therefore placed a great deal of effort into the development of creative methods that would ensure the successful involvement of users in the very early stages of prototype development.

For this purpose, new methods were used to allow for generation of new design solutions and evaluation of design solutions derived from the first phases of user involvement. Creative methods included theatre groups as well as specially designed focus groups using multimedia demonstrators. Use cases were thereby transformed into plays and scenes within a multimedia demonstrator which were then viewed and discussed within small user groups.

6.1 The Multimedia Demonstrator

Scenarios were presented to users via an animated movie. The multimedia demonstrator consists of Adobe Flash files that can be viewed in a web browser started via an html document. The demonstrator starts with a side-view presentation of a prototype SOPRANO house. Four rooms are visible: the living room, the kitchen, the bedroom and the bathroom. The entrance door is also visible. Each scenario takes place in one or more of these rooms. By selecting a room, the scenarios available for this room appear. The moderator can then choose from the various scenarios and the four available languages.

The multimedia demonstrator was applied in each of the design idea generation and design idea evaluation sessions. As a first step, the multimedia demonstrator was used to show participants the problem situation alone. After this the demonstrator was paused and people were asked whether they could identify with the presented problem situation. Participants were then encouraged to develop their own ideas in relation to the problem situation presented (technology-free ideas first, followed by technology-based ideas). After this part of the discussion the whole use case was presented using the multimedia-demonstrator.

The presentation of scenarios based on a multimedia demonstrator is a promising way to get detailed feedback of early stage prototypes. The overall system as well as individual pieces of technologies and interfaces can be experienced from the general to the more specific. As stated by Bierhoff and Panis (2008), this approach enables the stepwise integration of more and more concrete technology throughout the project runtime, which is a great advantage as the presentation setting can be adapted to the change in prototype sophistication.

6.2 The Theatre Method

Scenarios were also explored by using a theatre approach. The aim of the proposed theatre method was to find one conceptual design solution per scenario. Theatre methods can portray a situation in a very natural and immediate manner, which makes it easier to imagine and remember a scene. Plays are very suitable for activating memories and emotions of spectators (Bortz & Doring, 1995). The theatre method was conducted with a group of 8 participants in Newham (UK) by the partners, University of Liverpool and WRC in Dublin. The problem scenarios were acted out by a professional theatre group.

A discussion then took place, where the participants were asked to produce their own ideas about how to best support elderly people in the respective scenario from their point of view. These ideas were collected. In a second step the theatre group acted out the design solution. The participants were asked to give their feedback on the solution. Subsequently, there was a discussion in which a comparison was made between the design ideas generated by the participants and the design solutions developed by the project partners. Finally, the group decided on one design solution they felt was useful, acceptable and usable. Feedback from the participants can be visualised "on the fly" by playing the scene according to the ideas of the participants. The theatre method therefore allowed for very quick iteration and was very flexible to user input, since the moderator encouraged the participant to use target-oriented questions to develop their own ideas and to give feedback to solutions that were presented.

6.3 Results: Refinement of Scenarios and Generated Design Ideas as Input for Lab-Testing of Prototypes

Multimedia demonstrators and theatre groups turned out to be successful approaches for creating and evaluating design ideas. A total of 72 users were involved in 27 sessions conducted in 4 different countries in this phase of the research. Results coming from this cycle of user interaction revealed valuable refinements to each of the existing scenarios in relation to different design categories such as functionality, interaction sequence and modality. Some general requirements also came up during the discussions that are applicable to almost every scenario. Among these were that the system must be easy to operate, that the modality for reminding, informing, and alerting must be configurable, and that having control over the system is essential. Furthermore, reminders and notifications should not interrupt TV programmes and the system must take into account that users prefer not to leave TV in stand-by mode. Additionally, users always want to be informed about what has been done, e.g. that help is underway, appliances have been switched-off, a third person has been informed and so on.

Results from this cycle of user involvement would help technical designers to improve the prototypes. These prototypes were lab-tested at four sites with more than 50 users in the next stage of research. The lab-testing of prototypes focused on usability of the different SOPRANO components. The various methods applied ranged from observation based on common guidelines, execution of scenario tasks, user walk-through and questionnaires. These methods of classical usability testing were adapted to the needs of elderly people. For example, special emphasis was placed on giving the elderly users support when getting into the situation by giving them pictures. It was also important to remove the fear of failure by emphasizing that the system and not the user was being tested and to encourage open comments and gain the users' trust by stressing the importance of their role in the development process and listening to them carefully even when speaking off topic. Results were for used to revise the menu structure for the user interface, for example, or to facilitate the understanding of assignment of functionalities to remote control buttons (SOPRANO, 2008). Various components were tested: TV (including digital TV), speech generation, speech recognition, remote control, avatar, touch screen and web-based GUI. On the basis of the outcomes of this stage, components will be further improved.

The theatre method was experienced as particularly effective in terms of the expected advantages of the methods. The participants could easily identify with the actors and could imagine the problematic situation vividly. The elderly people were highly motivated to take part in discussions concerning the SOPRANO functionality. The theatre method was very effective in helping participants to visualise the use of the technology within the use case and to provide interactive feedback during the session. Furthermore, this method was experienced as very flexible as the interaction between the various participants (actors, researchers, and older people) was very dynamic. It greatly helped people participating in the creative process of idea generation, and is seen as providing many insights into a complex situation.

7 Conclusion

Requirements research within the SOPRANO project focused on increasing an understanding of the older person, exploring the situations, activities, cognitions, understandings and emotions that constitute an older person's experience and identifying key challenges to independent living. Applying multiple methods revealed valuable feedback from end users which could be applied to the technical development process, thus enabling the project team to further improve the different SOPRANO components. Formative evaluation during the design phase and evaluation of prototypes were both critically important for ensuring that the final system was usable and useful for its intended user population.

SOPRANO thus took takes an iterative approach by involving users in each of the phases of system development. The very first interaction with potential end users took place within the framework of elicitation of the requirements and consisted of focus groups and individual interviews in four EU countries to elicit user requirements in terms of key challenges to independence and initial ideas for possible technological solutions. These ideas were transformed into use cases and were then validated and further explored in a second stage of user research, utilizing multimedia mock-ups and theatre presentations to help users visualise technologies prior to prototyping. Feedback from users was used to refine use cases and finalise requirements. The next cycle of user interaction focused on testing usability of the different SOPRANO components and resulted in concrete design refinements of certain aspects of the SOPRANO components. The interconnection between the different stages of user involvement is presented in Figure 4 below.

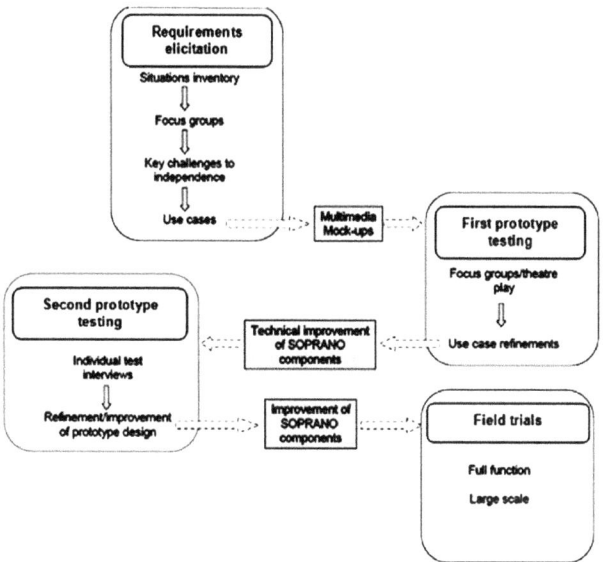

Figure 4: Iterative system development (Source: SOPRANO, 2007c)

SOPRANO has been highly innovative both in terms of its approach to the research and development process and in terms of the social and technical work being carried out. The state of the art approach for involving users typically involves the collection and analysis of user needs and requirements as the first step, followed by the development of prototypes by experts. By doing this, decisions about the conceptual design, i.e. what kind of functions are to be developed and the nature of the interaction, are made by experts. The prototypes based on those conceptual design ideas are then evaluated by users. The initial ideas are not based on the experiences of the users but on the experts. The user can intervene only through the fourth step – user based assessment. The E&AR approach was designed to change this situation. Instead of having experts lead the development of prototypes, a more user-centred design process was developed to involve users when developing the conceptual design and, later, the prototypes. Rather than users responding to the ideas of experts, experts should listen and respond to the input from older people. A key concern is to move away from a technology-push model to a more person-focused approach to ensure that technical solutions are sensitive to the real needs of older people.

Theatre methods and multimedia demonstrators also hold great promise for supporting idea generation. A group of multidisciplinary experts is used to match those ideas with what is achievable with today's technology. The dialogue between partners who come from a range of multidisciplinary backgrounds guarantee high acceptability by end-users in combination with technical feasibility.

Based on the evaluations of the SOPRANO system components it can be concluded that technology is not the solution to create a perfect home environment but it can make a useful contribution (Bierhoff & Panis, 2008). Overall satisfaction of end-users is dependent on the environment as a whole, including social contacts, for instance, what options are available for support and services, the location of the home and the medical problems people face. Therefore it is important to gain insight into all these aspects of the users participating in the research. As concluded by Bierhoff and Panis (2008), results also indicate that users allow technology into their homes; however not at any price. Involving them early on in the design process can result in greater acceptance and actual use of the technology. The participative approach developed and applied in the SOPRANO project allows us to drive the development of smart home technology based not only on what is technologically possible but also on how technology can help meet users' needs and requirements and thus lead to high acceptance and therefore innovation.

Bibliography

Bierhoff, I. & Panis, P., 2008. Active Involvement Of Older Users In The Design Process Of Smart Home Technology. Paper presented at the *6th International Conference of the International Society for Gerontology*. Pisa, Italy, 4-7 June 2008.

Burmester, M. & Machate, J., 2003. Creative Design of Interactive Products and Use of Usability Guidelines - a Contradiction? Human-Computer Interaction: Theory and Practice (Part 1), *Volume 1 of the Proceedings of HCI International 2003*, pp. 434-438. Lawrence Erlbaum.

Bortz, J. & Doring, N., 1995. *Forschungmethoden und Evaluation fur Sozialwissenschaftler*. Berlin: Springer.

Cuttler, C. & Graf, P., 2004. Individual differences in prospective memory. Paper presented at the *28th International Congress of Psychology*. Beijing, China, 8-13 August 2004.

Comyn, G., Olsson, S., Guenzler, R., Özcivelek, R., Zinnbauer, D. & Cabrera, M., 2006. User Needs in ICTResearch for Independent Living, with a Focus on Health Aspects. Report on a Joint DG JRC/IPTS-DGINFSO Workshop held in Brussels, 24-25 November 2005. [Online]. Available at: http://ec.europa

.eu/information_society/activities/health/docs/events/indep-living-nov2005/24-25nov-report-final-draft-june2006.pdf [Accessed 26 February 2010].

Demiris, G., Rantz, M. J., Aud, M. A., Marek, K. D., Tyrer, H. W., Skupic, M. & Hussam, A. A., 2004. Older adults attitudes towards and perceptions of 'smart home' technologies: a pilot study. *Medical Informatics*, 29 (2), pp. 87–94.

DIN EN ISO 13407, 1999. *Human-centred design processes for interactive systems*. Berlin:.Beuth.

European Commission, 2004. ISTAG Report on Experience and Application Research - Involving Users in the Development of Ambient Intelligence. Luxembourg: Office for Official Publications of the European Communities.

Katz, S., 1983. Assessing self-maintenance: Activities of daily living, mobility and instrumental activities of daily living. *JAGS*, 31 (12), pp. 721-726.

Kitwood, T., 1993. Towards a theory of dementia care: the interpersonal process. *Ageing & Society*, 13, pp. 51-67.

Lawton, M. P., 1990. Aging and performance of home tasks. *Human factors*, 32 (5), pp. 527-536.

McCreadie, C. & Tinker, A., 2005. The acceptability of assistive technology to older people. *Ageing and Society*, 25 (1), pp. 91-111.

Melenhorst, A-S., Rogers, W. A. & Bouwhuis, D. G., 2006. Older Adults' Motivated Choice for Technological Innovation: Evidence for Benefit-Driven Selectivity. *Psychology and Aging*, 21 (1), pp. 190–195.

Morell, R. W., Mayhorn, C. B. & Echt, K. V., 2004. Why Older Adults Use or Do Not Use the Internet. In: D. C. Burdick and S. Kwon, eds. *Gerotechnology: Research and Practice in Technology and Aging*. New York: Springer Publishing Company, pp. 71-85.

Müller, S., Santi, M. & Sixsmith, A., 2008. Eliciting user requirements for ambient assisted living: Results of the SOPRANO project. Paper presented at *The eChallenges e-2008 Conference*. Stockholm, Sweden, 22-24 October 2008.

Oswald, F., Wahl, H-W, Schilling, O., Nygren, C., Fänge, A., Sixsmith, A., Sixsmith, J., Széman, Z., Tomsone, S. & Iwarsson, S., 2007. Relationships Between Housing and Healthy Aging in Very Old Age. *The Gerontologist*, 47 (1), pp. 96-107.

Richardson, M., 2006. *Interruption events and sense-making processes: A narrative analysis of older people' relationships with computers*: PhD Doctorate. The University of Waikato.

Richardson M., Weaver, C. K. & Zorn, T. E., 2005. Getting on: older New Zealanders' perceptions of computing. *New Media & Society*, 7, pp. 219-245.

Rogers, W., Mayhorn, C. B. & Fisk, A. D., 2004. Technology in Everyday Life for Older Adults. In: D. C. Burdick and S. Kwon, eds. *Gerotechnology: Research and Practice in Technology and Aging*. New York: Springer Publishing Company, pp. 3-17.

Rosson, M. & Carroll, J., 2002. *Usability Engineering: Scenario-based development of Human-Computer Interaction*. San Francisco, CA: Morgan Kaufmann.

SeniorWatch - European Senior Watch Observatory and Inventory, 2002. *Older people and Information Society Technology*. [Online]. Available at: http://www.seniorwatch.de [Accessed 26 February 2010].

Sixsmith, A. & Sixsmith, J., 2000. Smart care technologies: Meeting whose needs? *Journal of telemedicine and telecare*, 6, 190-192.

Sixsmith, A. & Sixsmith, J., 2008. Ageing in place in the United Kingdom. *Ageing International*, 32 (3), pp. 219-235.

Sixsmith, A., Müller, S., Lull, F., Klein, M., Bierhoff, I., Delaney, S. & Savage, R., 2009. SOPRANO – An Ambient Assisted Living System for Supporting Older People at Home. Paper presented at the *7th International Conference on Smart Homes and Health Telematics*. Tours, France, 1-3 July 2009.

SOPRANO, 2007a. Deliverable D1.1.3: Review of social & cultural aspects.

SOPRANO, 2007b. Deliverable D1.2.2: SOPRANO Requirements Specification.

SOPRANO, 2007c. Deliverable D6.2.1: E&AR methodology for elderly needs collection and analysis.

SOPRANO, 2008. Deliverable D5.1.1: Results of tests relating to SOPRANO model & architecture, PART A.

Vimarlund, V & Olve, N. G., 2005. Economic analyses for ICT in elderly healthcare: questions and challenges. *Health informatics*, 11 (4), pp. 309-321.

Acknowledgements

SOPRANO (http://www.soprano-ip.org/) is an Integrated Project funded under the EU's FP6 IST programme Thematic Priority: 6.2.2: Ambient Assisted Living for the Ageing Society (IST – 2006 – 045212). The authors acknowledge the input and role of the SOPRANO consortium and would also like to thank the many people who volunteered in the various stages of user research described in this paper.

Alberta Contarello, Mauro Sarrica, Diego Romaioli

Ageing in a Broadband Society. An Exploration on ICTs, Emotional Experience and Social Well-being within a Social Representation Perspective

1 Introduction

In the last few years we have been involved in research on social representations of ICTs, mainly the internet and the mobile phone, on the one hand, and on social representation of ageing, on the other. Far as they might appear, the two topics are increasingly intertwined in the scholar debate (Chen, 2008). Both are facing very rapid changes and seem to constitute useful 'windows' on changes in contemporary society, mainly at the intersection between individual and social processes, the key node of social psychology.

Aim of the present contribution is to briefly outline the rationale and the main results of our previous studies and to present a new investigation centered on ageing, ICTs and social and psychological well-being, enquired within an intercultural perspective.

1.1 Social Representations of ICTs

Considering the spread of ICTs in the last decades an intriguing phenomenon at the intersection of knowing, thinking and acting in changing social environments, in previous studies we focused our attention on the internet and the mobile phone as social representations. Social representations are defined as shared structures of knowledge, loaded with 'warm' feelings and symbolic elements. Following the theory, an active co-construction takes place in the social-psychological arena when novelties of a theoretical, ideological and technological nature appear (Moscovici, 1961-1976). In the socio-dynamic extension of the theory, the resulting structures are governed by underlying principles along which pre-existing sociological and psychological systems operate their anchoring effects (Doise et al., 2003) Along this framework, we explored the representation of the internet and the mobile in relation to the human body in an Italian and a cross-national context (Contarello & Fortunati, 2006; Contarello et al. 2008). We also enquired the social representation of the mobile (Fortunati & Contarello, 2005) and its eventual transformation in the arch of a few years

(Contarello et al., 2007). Similar in the different studies, the research design was based on a quali-quantitative approach combining textual analyses and multivariate analyses with the aim to investigate the spiralling links between changes in social situations, psychological processes, and uses and practices, well represented in Flick's schematization of the reference theory (Figure 1).

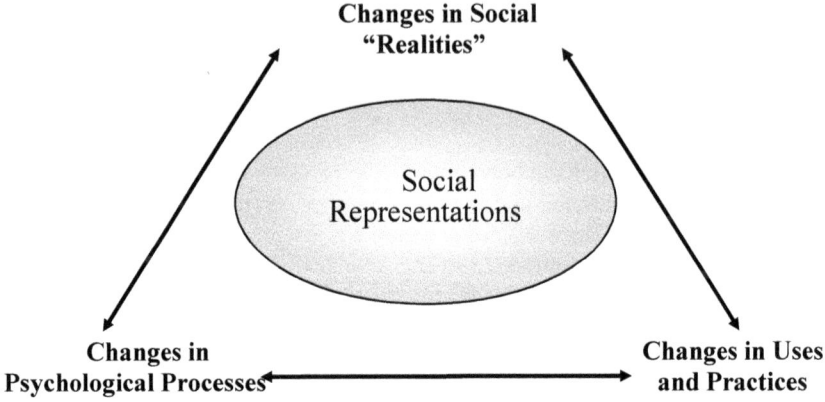

Figure 1: The Space in Between (Source: adapted from Flick et al., 2002)

In a similar vein we enquired the social representation of the internet in relation to different forms and levels of practice by its users and to self-reported social well-being in young participants (Contarello & Sarrica, 2007). We spend some words on this study as it constitutes the format of the first part of the present research. One hundred university students were invited to answer a free association task to the stimulus-word "the internet", a social well-being scale (Keyes, 1996) and a questionnaire on practices and uses of the web. Correspondence analysis carried out on the associated terms gave rise to three main underlying principles which we interpreted as an inwards vs. outwards perspective, space vs. time extension, and function (what the internet is for) vs. experience (how the internet is experienced). The design variables - uses and practices, attitudes, social well-being (both in general and after the advent of the internet) - proved to enter into the picture in several ways. As regards social well-being, we found that, overall, among the participants – already showing positive levels of social well-being – the perception of change in their life, since the internet entered it, was one of improvement. But also the five dimensions composing the measure of well-being – social integration, social acceptance, social contribution, social actualization, social coherence – showed to be related with views of the internet. For instance, those who reported greater trust in people since they came familiar with the internet, mention a dynamic acceleration in study, work and friendly bonds.

1.2 Social Representations of Ageing

Social Psychology has provided important contributions to the study of ageing as a social issue, both from a social cognition perspective – in terms of social categorization and social identity (Brewer & Lui, 1989), stereotypes and their threat (Hummert, 1990), "ageism" (Giles & Reid, 2005) – and, more recently, from a social representation perspective. This theoretical framework, as above mentioned, invites to take into account, jointly, psychological processes and social dynamics, thus providing a suitable arena in which to explore and deepen our knowledge of social-psychological questions. Ageing has been thus read and interpreted through this theoretical lens (Contarello et al., n.d.; Coudin & Paicheler, 2002; Jodelet, 2008; Jesuino, 2008; Martins et al., n.d.), also from a cross-cultural (Liu et al., 2003; Journal of Cross-Cultural Gerontology) and intercultural stance (Contarello et al., n.d.).

In our studies, we found support to a rich and varied social representation of ageing both in young and elderly Italian respondents, with key themes of potentialities and weaknesses at the forefront. We also found support to the well-known inner-outer perspective characterizing groups: the youth linking ageing with illness, the elderly considering the two topics not necessarily correlated (Gastaldi & Contarello, 2006). Addressing the pivotal role played at a societal level by mass-media communication, we also explored via archive research the representation of ageing in Italian weekly magazines and enquired, via focus groups with elderly people, their view of the resulting picture. The results show a scarce presence of the elderly in printed advertising and an unrewarding and/or unrealistic one. In the words of the respondents, either too active or too passive characters are depicted, that would foster stereotypical images, which – in a mirror like effect – would damage the life of the elderly within society.

Not very many facets of new theories on ageing stressing the positive sides of the process - Successful Ageing (Baltes & Baltes, 1990), Positive Aging (Gergen & Gergen, 2000), Active Ageing (WHO) – are easily encountered in our respondents' productions (be they collected with interviews, or questionnaires, or free association tasks). However, several positive features appear here and there, although being seldom integrated into a stable structure.

More recently, a qualitative study has been carried out exploring social representations of ageing in an intercultural framework, via interviews with Italian and Brazilian respondents – in Padova, Italy and Florianopolis, Santa Catarina, Brazil (Contarello et al., n.d). Asked to narrate and discuss the image of an aged person and views on ageing held in one's own and in the other cultural context, young, adult and elderly respondents, evoke mostly a scenario of poverty and social exclusion, but also consider positive features and enhance the process of change which is characterizing ageing at a global level. More optimism appear in the Brazilian answers, with the hope of an improved situation for the elderly linked with an increased economic power, while more pessimism cross the Italian answers connected with the acknowledgment of a relative independence between economic and cultural resources. In both cultural contexts, good antidotes to the dark features of ageing are indicated in interpersonal relations and social cohesion, with a curious inversion: the respondents (mainly younger ones) ascribe to the other Country the buffering power of family and traditional bonds.

In the foreground, a certain reluctance remains to endorse a single, "objectified", view of ageing and of the aged person, which result quite different, according as social position, income, health condition, previous and present activities. As regards the respondents, more than age, what seems at stake is generation, with different concerns and priorities in different age cohorts.

2 Ageing, ICTs, Social Well-Being

The research we would like to present is composed of two parts.

The first inquires representations of ICTs, social well-being and emotional experience in elderly participants of different age. It has been carried out with a paper&pencil instrument, including a free associations task, and processed through quali-quantitative analysis, in a similar way as previously reported. Particular attention has been given to the content, the attitude and the representation field of ICTs, internet and the mobile phone - as well as to the relation between views of ICTs, uses and practices and reported social well-being.

The second deepens our focus of attention with interviews with 'first generation' elderly users.

As data are still under process, we focus here our attention on a first data collection and analysis which we present in terms of pre-research.

2.1 Aims

As anticipated, our main aim was to explore how the internet and the mobile phone – seen as emblematic of ICTs – are perceived and constructed in relation to different forms and levels of practice by individuals and how these social constructions are connected with self-reported social well-being. A particular interest was devoted to the age range of potential users, so we considered separately three age levels close to the ones which in the literature on ageing are referred to as young elderly, elderly, old or great elderly.

2.2 Participants

The participants were one hundred elderly people subdivided into three groups (65-69 N54, 70-74 N21, 75-84 N24), women N55 and men N45 living in Northeast Italy. Data were collected in 2006[1].

2.3 Research Design and Procedure

The research design was very similar to the study on internet and well-being previously presented: we explored content, attitude and representation field of ICTs, particularly the internet and the mobile, held by the respondents. First, we monitored their social well-being with an adaptation of Keyes's (1998) scale (reduced to 22 subscales instead of 33) designed to measure social integration, acceptance, contribution, actualization and coherence, as self-reported by the respondents. Second we collected quali-quantitative data via free associations, semantic differential scales and a short questionnaire, mainly on uses and practices. The stimulus-words for the free association task were: internet, mobile, myself, elderly people using ICTs. Third we measured again (separately, in a shorter version of the scale) perceived social well-being after the advent of the internet and the mobile.

The participants also answered questions monitoring ownership, familiarity with the internet, the computer and the mobile phone, as well as gender, age, education, residence, social networks and status of relationships with others.

2.4 Data Analysis

Textual data were analysed with the help of the Spad.T package (Lebart et al., 1989): after the usual pre-treatment of them, reducing the synonyms and fixing an inclusion frequency threshold, they were treated through correspondence factor analysis (Benzécri et al., 1973; Clausen, 1998).

The two participants x words matrixes referring separately to the stimuli internet and mobile phone were analysed. This helped us to explore the emerging representational field and the role played by the illustrative variables along the organising principles which define it[2].

---

1   We gratefully acknowledge Francesca Michelin and Silvia Prati's help in the phases of data collection and analysis.
2   For more details, see Contarello A. & Sarrica M., 2007. Icts, Social Thinking And Subjective Well-Being. The Internet And Its Representations In Everyday Life. *Computers In Human Behaviour*, 23, pp.1016-32.

2.5 Results

As regards social well-being, the participants reported a medium/high level of well-being in terms of social integration and contribution and a medium level of social acceptance.

2.5.1 The Internet

More than one third of the respondents (38%) reported to own a PC, more men (48.9%) than women (29.1%). This was more diffused in the young elderly (50%) than in old ones (21%). One third (32%) owned an internet connection, again with greater diffusion in women and younger respondents. Internet was mostly used to search for information and navigate (92%). Only 9% of the participants owned an email address, more frequently the younger ones (between 65-69, 60%).

Analysing the content of the representation from free associations, we found a positive and pragmatic view of the internet, as a source of progress, usefulness, information, but also critical aspects with terms such as taboo, something unknown, complex and dangerous.

As regards their attitude, the respondents reported a positive view of the internet both in terms of general evaluation (it is desirable, agreeable, pleasant, entertaining, not harmful) and of usefulness, but they also judged it difficult and artificial.

The representation field appeared somehow poor, with scarce associations and re-launches (cf. Table 1). The first factor opposes an evaluation in negative/ironic terms - *a container limiting thinking* - to a straight reference to the device supporting the web – *computer*. The first view is mostly held by young elderly, with medium/high familiarity with the technologies but with a lower evaluation of them, with medium (acceptance) or low (integration) levels of social well-being, while the second by old elderly with minimal or no familiarity with the devices. The following factors were interpreted by us as *complexity* vs. *usefulness*, *difficulty* vs. *taboo*, *progress* vs. *risk*, *novelty for young people* vs. *challenging informational device*. Table 1 illustrates, as an example, the second factor with the positioning of age, devices evaluation and perceived social well-being along the two poles of the dimension.

	COMPLEXITY		+	USEFULNESS	
	A.C.	R.C.		A.C.	R.C.
Difficult to use	23.3	0.26	Computer	14.3	0.19
Taboo	22.2	0.30	Informative	7.0	0.12
Complex	12.4	0.26	Communication	4.5	0.09
I do not use it	4.6	0.10			

75-84
Low internet evaluation

65-69
Average Social Integration
High internet evaluation

Table 1: The Internet. Correspondence Analysis. Second Factor (Source: the authors)

Comparing these results with those obtained in the previously mentioned study conducted with university students, we found considerable differences: for the youth the internet appears to be a tool both for information and communication (with associations such as computer, science, speed, pornography, fun), while for the elderly it is mostly an instrument, useful as informative for the young elderly, far from one's own terrain for the great elderly.

The general view on the technologies in relation to social well-being, however, is far from negative. On the whole, in the answers to the question regarding social well-being after the internet entered everyday life, this appeared to have improved, not dramatically but significantly, the perception of the respondents' social well-being (Table 2).

Social Well-being 'after the internet'	Mean
Social Integration	3.26**
Social Acceptance	3.18*
Social Contribution	3.30**
Social Actualization	4.04**
Social Coherence	3.51**

Table 2 Differences from the average score (* p>.05; ** p>.01) (Source: the authors)

2.5.2 The Mobile

A different picture regards the mobile phone. A great deal of the respondents (78%) reported to own a mobile, with smaller difference between men (86.7%) and women (70.6%). This was again more diffused in the young elderly (90.7%) than in old ones (54.3%). Only 20% of the participants reported to use texts (Sms), mostly the younger ones (26.9%).

Analysing the content of the representation from free associations we found again a positive and pragmatic view of the device, as a source of usefulness, convenience, help in cases of emergency, but also critical aspects with terms such as annoyance, not strictly necessary, dangerous as regards privacy.

Coming to the attitude, the respondents reported a positive view of the mobile both in terms of general evaluation and of usefulness.

The representation field appeared to be based on five main dimensions. We interpreted the first as *convenience* vs. *no necessary need* and the fifth as *emergency* vs. *valid but annoying device*. The intermediate three factors always involve *family* on one pole counterbalanced, on the other, by views of the mobile as *useless convenience, annoying device, valid instrument*. Reported social well-being enters into the picture showing that higher social integration and contribution go hand in hand with linking the mobile to emergencies, while lower social acceptance is connected with recognition of convenience but not necessarily of usefulness of the device. Age plays a more limited role than for the internet, presenting an effect only on the fourth factor where particularly the elderly in their early seventies associate positive nuances to the mobile, considering it as a valid instrument.

Overall, again, the general view on the mobile in relation to social well-being results positive. Social well-being, after the mobile entered everyday life, appears to have improved, not dramatically but significantly, the perception of the respondents' social well-being (Table 3)

Social Well-being 'after the mobile'	Mean
Social Integration	3.17*
Social Acceptance	3.19*
Social Contribution	3.37**
Social Actualization	4.09**
Social Coherence	3.48**

Table 3: *Differences from the average score (* p>.05; ** p>.01) (Source: the author)*

3 Concluding Notes in Progress

Provisional as they might be, these preliminary notes show how important, in the study of the relation between representations of ICTs and social psychological processes, is to take into account different components. Not age in itself, but a rich variety of variables have to be considered analysing the relationship between ageing and ICTs. In this, we align with the conclusions by Maria Sourbati, in her qualitative study with elderly users and non users in sheltered houses in London, that "policy development must abandon its technology-centric focus and take a broader, interdisciplinary perspective on the diversity of older users, their social material and cultural circumstances, their needs and wishes, and their everyday practices of media (and) service use" (2000, p.102).

This view is also supported by results obtained through qualitative and quantitative research in various cultural contexts. Research carried out in Great Britain, for instance, supported by independent French data, stress the importance of values, habits and tastes of generations (Haddon, 2004). More than to age in itself, these might be linked to an age/cohort effect, based on the differential diffusion of domestic information and communication technology along time (Gilleard & Higgs, 2008). But great relevance appear to be played by social psychological processes linked with practical aims and social dynamics. Qualitative research helped and might help to better understand the issue of ambivalence towards ICTs which we also encountered in our data[3], as well as the nature and meaning of elder participants and non-participants in cyberspace[4]. Research is growing along these lines, as well as on the relationships between ageing, the internet and well-being (Amichai-Hamburger & Barak, 2009). We ourselves have been extending our research with interviews with elderly users and non-users of ICTs in different cultural contexts, actually in progress. We hope thus to contribute to a deeper social psychological understanding of the situated relationships between ageing and media in a broadband society.

---

[3] For an interesting study in England and South Wales: Selwyn N., 2004. The information Aged: A Qualitative Study of Older Adults' Use of Information and Communications Technology. *Journal of Aging Studies*, 18, pp.369-384

[4] For an interesting study with Israeli retirees: Blit-Cohen E. & Litwin H., 2004. Elder Participation in Cyberspace: A Qualitative Analysis of Israeli Retirees. *Journal of Aging Studies*. 18, pp.385-398.

Bibliography

Amichai-Hamburger, Y. ed., 2009. *Technology and Psychological Well-Being*. Cambridge: Cambridge University Press.

Amichai-Hamburger, Y. & Barak, A., 2009. Internet and Well-Being. In: Y. Amichai-Hamburger, ed., *Technology and Psychological Well-Being*. Cambridge: Cambridge University Press.

Baltes, P.B. & Baltes, M.M., 1990. Psychological Perspectives on Successful Ageing: The Model of Selective Optimization with Compensation. In P.B. Baltes & M.M. Baltes, eds., *Successful Ageing. Perspectives from the Behavioral Sciences*. Cambridge: Cambridge University Press.

Benzécri, J.P. et al., 1973. *L'Analyse de Données. L'Analyse des Correspondances* (vol.2). Paris: Dunod.

Blit-Cohen, E. & Litwin, H., 2004. Elder Participation in Cyberspace: A Qualitative Analysis of Israeli Retirees. *Journal of Aging Studies*. 18 (4), pp.385-398.

Brewer, M.B. & Lui, L., 1989. The Primacy of Age and Sex in the Structure of Person Categories. *Social Cognition*.7 (3), pp.262-274.

Chen, H., 2008. Introduction to Special Section on Aging and the Internet. *Ageing International*, 32 (1), pp.1-2.

Clausen, S.E., 1998. *Applied Correspondence Analysis*. London: Sage.

Contarello, A., Contarello, L. & Bonetto, R., 2008. With the Eyes of a Bee. An Incoming Vision. In B. Sapio et al., ed., *The Good, the Bad and the Unexpected. The Users and the Future of Information and Communication Technologies* (Cost Action 298 "Participation in the Broadband Society"). Brussels: Cost Office.

Contarello, A., Bonetto, R., Romaioli, D. & Wachelke, J. *Invecchiamento e intercultura*. Forthcoming.

Contarello, A., Bonetto, R., Romaioli, D. & Wachelke, J. *L'invecchiamento tra dinamiche intergenerazionali ed interculturali*. Forthcoming.

Contarello, A. & Fortunati, L., 2006. ICTS and the Human Body: A Social Representation Approach. In P. Law, L. Fortunati & S.Yang, eds., *New Technologies in Global Societies*. Singapore: World Scientific Publisher, pp.51-74.

Contarello, A., Fortunati, L. & Sarrica, M., 2007. Social Thinking and the Mobile Phone. A Study of Social Change with the Diffusion of Mobile Phones, Using a Social Representations Framework. *Continuum*, 21 (2), pp.149-162.

Contarello, A. et al. 2008. ICTs and the Human Body: An Empirical Study in Five Countries. In: E. Loos, E. Mantemeijer & L. Haddon ed. *The Social Dynamics of Information and Communication Technology*. Aldershot Hampshire: Ashgate, 2008, pp.25-38.

Contarello, A., Romaioli, D. & Bonetto, R., Ageing And Generations. A Social Representation Approach. In M. Lopez, F. Mendez e A. Moreira, eds., *Saude, Educacao e Representacions Sociales*. Evora: Formasau, pp. 89-104.

Contarello, A. & Sarrica, M., 2007. ICTs, Social Thinking and Subjective Well-Being. The Internet and Its Representations in Everyday Life. *Computers in Human Behaviour*, 23, pp.1016-32.

Coudin, G. & Paicheler, G., 2002. *Santé et viellissement*. Paris: Armand Colin.

Doise, W., Clemence, A. & Lorenzi Cioldi, F., 2003. *The Quantitative Analysis of Social Representations*. Hemel Hempstead: Harvester Wheatsheaf.

Flick, U., 1998. The Social Construction of Individual and Public Health: Contributions of Social Representations Theory to a Social Science of Health. *Social Science Information*, 37 (4), pp.639-662.

Flick, U., Fischer, C., Schwartz, F.W. & Walter, U., 2002. Social Representations of Health Held by Health Professionals: the Case of General Practitioners and Home-care Nurses. *Social Science Information*, 41 (4), pp. 581-602.

Fortunati, L. & Contarello, A., 2005. Social Representation of the Mobile: An Italian Study. In Shin Dong Kim, ed., *When Mobile Came. The Cultural and Social Impact of Mobile Communication*. Seoul: Communication Books, pp.45-61.

Gastaldi, A. & Contarello, A., 2006. Una questione di età. *Ricerche di Psicologia*, 20 (4), pp.7-22.

Gergen, K.J. & Gergen, M.M., 2000. The New Aging: Self Construction and Social Values. In K.W. Schaie, ed., *Social Structures and Aging*. New York: Springer.

Giles, H. & Reid, S.R., 2005. Ageism across the Lifespan: Towards a Self-Categorization Model of Age. *Journal of Social Issues*, 61 (2), pp.389-404.

Gilleard, C. & Higgs, P., 2008. Internet Use and the Digital Divide in the English Longitudinal Study of Ageing. *European Journal of Ageing*, 5 (3), pp.233-239.

Haddon, L., 2004. *Information and Communication Technologies in Everyday Life*. Oxford: Berg.

Hummert, M.L., 1990. Multiple Stereotypes of Elderly and Young Adults: A Comparison of Structure and Evaluation. *Psychology and Aging*, 5 (2), pp.182-193.

Jesuino, J., 2008. Conferencia Abertura. *I Coloquio Luso-Brasileiro sobre Saude, Educacao e Rpresentacoes Sociais*. Evora, Portugal, March 12-15th 2008.

Jodelet, D., 2008. Contributo das representacoes sociais para o dominio da saude e da velhice. In M. Lopez, F. Mendez e A. Moreira, eds., *Saude, Educacao e Representacions Sociales*. Evora: Formasau.

Keyes, C.L.M., 1996. Social Well-being. *Social Psychology Quarterly*, 61 (2), pp.121-140.

Lebart, L. ., Morineau A., Becue M., Haeusler, L., 1989. *Système Portable pour l'Analyse des Données Textuelles* (SPAD.T). Paris: Cisia.

Liu, J.H., Ng, S.H., Loong, C., Gee, S. & Weatherall, A., 2003. Cultural Stereotypes and Social Representations of Elders from Chinese and European Perspectives. *Journal of Cross-Cultural Gerontology*, 18 (2), pp.149-168.

Martins, C. M., Camargo, B.V. & Biasus, F., 2009, Representações Sociais do Idoso e da Velhice de Diferentes Gerações. *Universitas Psychologica*, 8 (3), pp.831-847.

Moscovici, S., 1961-1976. *La psychanalyse, son image et son public*. Presses Universitaires de France, Paris.

Selwyn, N., 2004. The information Aged: A Qualitative Study of Older Adults' Use of Information and Communications Technology. *Journal of Aging Studies*, 18 (4), pp.369-384.

Sourbati, M., 2008. On Older People, Internet Access and Electronic Serivice Delivery: A Study of Sheltered Homes. In: E. Loos, E. Mante-Meijer & L. Haddon ed. *The Social Dynamics of Information and Communication Technology*. Aldershot Hampshire: Ashgate.

World Health Organization, 2002. Active Ageing. A policy Framework. Available at: http://www.who.int/hpr/ageing/ActiveAgeingPolicyFrame.pdf. [Accessed June 2009]

Eugène Loos

Generational use of new media and the (ir)relevance of age

1 Introduction

The use of new media in our information society is constantly increasing, as is the number of older people. In the year 2000, the Council of the European Union and the Commission of the European Communities presented an eEurope Action Plan entitled 'An Information Society For All' which set out three main objectives: the realization of a cheaper, faster, secure Internet, investment in people and skills, and encouragement to expand the use of the Internet. The second objective specifically stated that 'the Lisbon European Council recognised that special attention should be given to disabled people and fight against *info-exclusion*. (…) As government services and important public information become increasingly available on-line, ensuring access to government websites for all citizens becomes as important as ensuring access to public buildings.'

It is interesting to note that, while the e-Europe Action Plan made explicit mention of disabled people, it failed to address the issue of older citizens. In light of the growing number of older users in our information society, this is a group whose concerns also merit attention. The supply of digital information through websites and the like must be available to older generations, so that they have guaranteed access to the digital information sources provided by public and private organizations offering products and services they need. Some researchers argue that there is a widening generational *digital gap* between those people who are able to use new media and those who are not. It was Prensky (2001, pp. 1-2) who coined the notions of *digital natives* and digital *immigrants*. From an educational point of view, he considers students to be *digital natives* because they are 'all native speakers of the digital language of computers, video games and the Internet. So what does that make the rest of us? Those of us who were not born into the digital world but have, at some point later in our lives, become fascinated by and adopted many or most aspects of the new technology are, and always will be compared to them, Digital Immigrants'.

Do they really exist, these *digital natives*, who are identified as the generations born after 1990, who have grown up with new media? And is there really an older generation of *digital immigrants* playing catch-up by trying to learn how to use new media? Other researchers (Lenhart & Horrigan, 2003) take a different perspective. They introduced the notion of a *digital spectrum*, which acknowledges that people use new media to varying degrees, depending not only on age but also on factors such as gender, educational background and frequency of internet use.

If we want the supply of digital information through websites and the like to be readily available to older generations so that they are guaranteed access to the digital information sources provided by public and private organizations and the much-needed products and services offered there, we need to gain insight into their navigation behaviour. This paper therefore presents the results of an explorative case-study which focuses on the question of whether older people do indeed navigate websites differently from younger people. Or are the differences within this group arising from such factors as gender, educational background and frequency of internet use greater than the differences to be found between younger and older people?

2 Characteristics of Older Internet Users: A Quick Scan

Are there studies which offer insight into the way older people navigate websites and the factors which help or hinder their ability to gain access to such digital information? Let us first have a look at the field of research into aging. The Handbook of Communication and Aging Research, edited by Nussbaum & Coupland in 2004, offers a compilation of the research carried out in this field over the past three decades. The book is divided into seven sections, each dealing with a particular aspect of study. The first section deals with the experience of aging, the second with language, culture and social aging. The third section examines the communicative construction of relationships in later life, the fourth organisational communication, and the fifth political and mass communication. Section six addresses health communication and lastly, section seven discusses senior adult communication. In almost none of these sections is any attention paid to the use of digital information sources by older people: the only exception is in section four, where, in chapter 13 under the title 'Marketing to Older Adults', Balazs (2004) discusses communication with older people. However, rather than dealing with how older people browse websites in search of information, her focus is on selling products and services.

The proceedings of the 4th International Conference on Universal Access in Human-Computer Interaction in Beijing in July 2007, edited by Stephanidis, perhaps offer more insight into the way older people looking for information navigate websites. Part IV of 'Understanding Diversity: Age' indeed presents several studies focusing on older people using new media and how this can make their lives more comfortable. One of the studies entitled *Older Adults and the Web: Lessons learned from Eye-Tracking*, by Tullis, put forward empirical research results about differences between younger and older U.S. users in the way they scan web pages: 'An eye-tracking study of a prototype website was conducted with 10 younger adults (ages 20-39) and 10 older adults (ages 50-69) to determine if there are differences in how they scan webpages. They performed the same task on the website. On the average, the older adults spent 42% more time looking at the content of the pages than did the younger adults. They also spent 51% more time looking at the navigation areas. The pattern of fixations on almost all pages showed that the older adults looked at more parts of the page then did the younger adults. (...) One thing we did not see was any difference in the likelihood of older and younger users to view page content *below the fold* (i.e., that they had to scroll to view)' (Tullis, 2007, pp. 1030, 1038).

This study shows potentially significant patterns which may be validated by future comparable research. An example of similar empirical research is that done by Houtepen who conducted an explorative eye-tracking study on 13 younger users (18-25 years old) and 7 older users (over 50) in the Netherlands in 2007. As in Tullis' study, the subjects were asked to perform a specific task, in this case, finding health care information. Two main points emerged from the study: the older users took more time to fulfil the task (nearly 6 minutes compared to the 2.5 minutes taken by the younger users); the older subjects read more and made less use of the website's search box facility.

Both Tullis' study and Houtepen's research show that older users need more time and follow a different reading pattern. A measurement study conducted by Pernice & Nielsen (2002) on 20 seniors and a control croup of 20 users aged between 21 and 55 using three websites and a Web-wide task confirms the differences in time needed for task completion: 12:33 minutes for the seniors versus 7:14 minutes for the younger control group. They also offer an explanation for this difference: 'Websites tend to be produced by young designers, who often assume that all users have perfect vision and motor control, and know everything about the Web. These assumptions are rarely upheld, even when the users are not seniors. However, as indicated by our usability metrics, seniors are hurt more by usability problems than younger users. Among the obvious physical attributes often affected by human aging process are eyesight, precision of movement, and memory.' Pernice & Nielsen (2002, p. 4).

The studies conducted by Tullis (2007), Houtepen (2007) and Pernice & Nielsen (2002), as well as the reviews by Chisnell & Redish (2004, 2005) and Andrew (2008), offer insight into differences related to time on task and reading patterns between younger and older users. However, it should be borne in mind that the studies involved a limited number of participants, which would point to the need for more research on more users. The studies also only focused on age, omitting to take into account the role of factors such as gender, educational background and frequency of internet use. It is therefore unclear whether differences within an age group are greater than the differences between younger and older people.

3 Explorative Case-study: Research Questions and Methodology

What is needed is research based on empirical studies with larger groups of older and younger generations which take into account the role of factors like gender, educational background and frequency of internet use. If we conduct eye-tracking studies with larger groups and pay attention to these factors and not only to age we will gain a better understanding of the differences and similarities related to navigation behaviour.

The question is how to set up and conduct such a study. In this paper, I make use of the data from an eye-tracking study carried out among 29 younger and 29 older users (respectively about 21 years old and 65 years and older) in the Netherlands in April 2009. For a complete overview of all the results of this study, see Loos & Mante-Meijer (2009). Compared to earlier empirical research conducted in this field, a relatively large number of participants were recruited to our empirical study. The number of 2 x 29 participants far exceeds the minimum of 8 participants per user type in usability tests specified by the NIST CIF (Wichansky (2000) in Goldberg & Wichansky, 2003, p.512). However, this number is still not large and the possibility of distortion remains. It is therefore accurate to characterise this case study as an exploratory one, in which we demonstrate trends rather than significant relationships. For this reason, no standard deviations and p-values are included.

Older people often are considered to be a more diverse group than younger people. Bouma (2000, p.68) for example explains that 'Education and job specialization have been rising all through the 20th century, and the new generations of older citizens have learned to be both assertive and active. It is certain that they will be a heterogeneous group, since cumulative life experiences vary so much more than among young adults.' I therefore also paid attention to the effect of gender, educational background and frequency of internet use (daily or otherwise) on the navigation behaviour of older people.

User group	N
All users	58
All older users	29
All younger users	29
All female users	28
All male users	30
All younger female users	14
All younger male users	15
All older female users	14
All older male users	15
All older users with higher education	19
All older users without higher education	10
All older users using internet daily	18
All older user not using internet daily	11

Table 1: Different user groups. (Source: Loos & Mante-Meijer (2009, p.46))

Heatmaps and gaze plots are used to show the output of the eye-tracking instrument which measures the eye-movements of the different user groups. Heatmaps use different colours to show how intensely navigation areas are visited based on the number of fixations by individual users or groups of users (red for high, yellow for moderate and green for low intensity). Gaze plots provide insight into the eye movements, or saccades, of individual users by presenting the order (numbers in circles) and duration of gaze fixations (the longer the gaze fixations the bigger the circle). Red-white demarcations show where users have clicked. See McElhal (2007) for more information about eye-tracking.

The users performed a search task on the website of ANBO, a Dutch organization for senior citizens. The users had to find information about discounts related to health insurance which could be found on a specific web page of the site (Loos & Mante-Meijer (2009) for more information). Their navigation behaviour, i.e. their reading patterns and use of search box, was then analyzed, with particular attention being paid to effectiveness (whether or not the search task was completed successfully within 5 minutes), efficiency (how long they took to fulfil their search task) and user satisfaction (usability ranking); see also Frøkjaer, Herzum & Hornbaek (2000) and Johnson & Kent (2007).

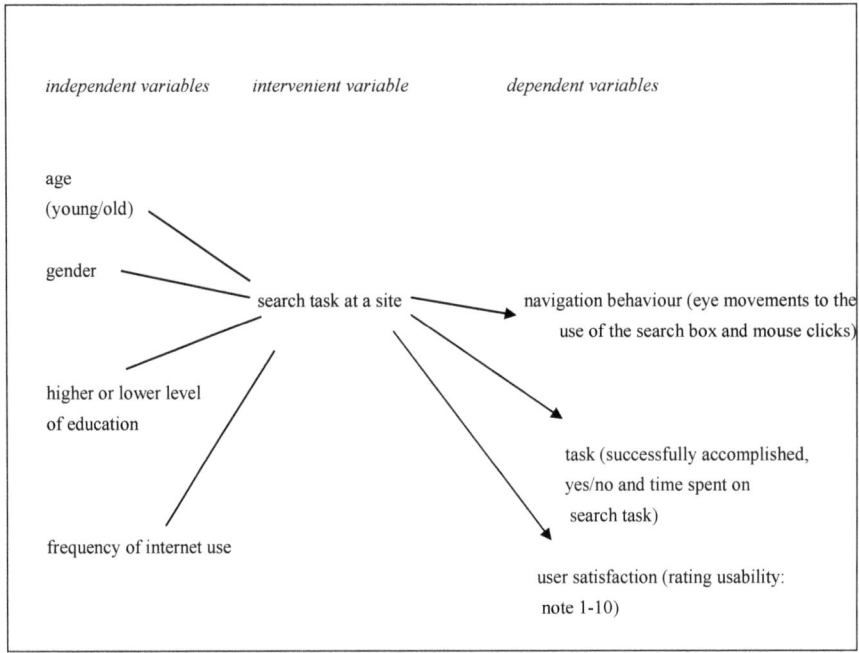

Figure 1: Research design (Source: Loos & Mante-Meijer (2009, p.48))

4 Results

4.1 Use of the Search Box

The heatmaps show that the majority of users made no use of the search box during the search task (Loos & Mante-Meijer, 2009, pp. 53-54). Table 2 confirms this result: only 13.8% of users in the older and 13.8% of the younger age group used the search box.
This result fails to confirm the findings of Houtepen's study (see section 2), which showed that older people used the search box less frequently than did younger users. A possible explanation is that the search task in my eye-tracking study was rather easy to perform, so most users apparently had no need to use the search box.
Other significant differences between user groups are presented in Table 2:
The percentage of the female users making use of the search box was higher than that of the male users: 17.9% versus 10%.

264

Only 6.7% of the older male users made use of the search box compared with 21.4% of the older female users.

Older people with a lower level of education used the search box more often than did older people with a higher level of education: 20% against 10.5%.

Older people making daily use of the internet used the search box in 22.2 % of the cases; of the group who did not use the internet daily, not a single person made use of the search box.

User Group	Search box used during search task		Search box *not* used during search task	
	N	%	N	%
All users	8	13.8	50	86.2
All older users	4	13.8	25	86.2
All younger users	4	13.8	25	86.2
All female users	5	17.9	23	82.1
All male users	3	10	27	90
All younger female users	2	14.3	12	85.7
All younger male users	2	13.3	13	86.7
All older female users	3	21.4	11	78.6
All older male users	1	6.7	14	93.3
All older users with higher education	2	10.5	17	89.5
All older users without higher education	2	20	8	80
All older users using internet daily	4	22.2	14	77.8
All older users not using internet daily	0	0	11	100

Table 2: Use of the search box (Source: Loos & Mante-Meijer (2009, p.46))

4.2 Navigation Areas

Loos & Mante-Meijer (2009, p. 57) argue that the navigation patterns of older and younger users seem to be different. Though many in both the younger and the older group of users looked at the right place to click to arrive at the web page containing the information they were looking for, the red area on the older users' heatmap is much larger than on the younger users' heatmap. This confirms Tullis' finding that older people examine navigation areas more intensely than do younger people. Another difference is that more older users than younger users look during a longer period at the wrong place to click. So, at first glance, the navigation patterns of older people appear to differ from those of younger people.

However, if we compare the navigation patterns of older people using the internet daily with those of the younger age group, these patterns are in fact not as dissimilar as first thought (Loos & Mante-Meijer, 2009, p. 57 and Loos, forthcoming).

This would seem to imply that the frequency of internet use impacts more heavily on our navigation patterns than does age.

4.3 Gaze Plots

I obtained the gaze plots of 12 older female users, 12 older male users, 10 younger female users and 13 younger male users looking at the homepage of the ANBO website (Loos & Mante-Meijer, 2009, p. 58). The gaze plots show users in each of these groups who needed a low (up to 40) or high number of saccades (over 40) to get to the next webpage where the information could be found (Loos & Mante-Meijer, 2009, pp. 58-59) After analysis of the gaze plots, it appeared that younger male users needed the lowest number of saccades to reach the webpage where the information could be found. Only 15% needed more than 40 saccades versus 33% of all younger female users and older male users, and 42% of older female users. The number of saccades does not necessarily predict effectiveness, efficiency and user satisfaction. Analysis of this can be found in the next section.

4.4 Effectiveness, Efficiency and User Satisfaction

4.4.1 Effectiveness (Search Task (not) Successfully Accomplished within 5 Minutes)

User Group	Search task successfully accomplished		Search task *not* successfully accomplished	
	N	%	N	%
All users	51	87.9	7	12.1
All older users	24	82.8	5	17.2
All younger users	27	93.1	2	6.9
All female users	23	82.1	5	17.9
All male users	28	93.3	2	6.7
All younger female users	12	85.7	2	14.3
All younger male users	15	100	0	0
All older female users	11	78.6	3	21.4
All older male users	13	86.7	2	13.3
All older users with higher education	16	84.2	3	15.8
All older users without higher education	8	80	2	20
All older users using internet daily	16	88.9	2	11.1
All older users not using internet daily	8	72.7	3	27.3

Table 3: Effectiveness (Source: Loos & Mante-Meijer (2009, p.61))

Table 3 shows clearly that most users (87.9%) accomplished the search task within the 5 minute time limit, although some differences were apparent between user groups:

82.8 % of all older users accomplished the search task successfully versus 93.1% of all younger users.

82.1% of all female users accomplished the search task successfully versus 93.3% of all male users.

All younger male users succeeded in accomplishing the search task successfully versus 85.7% of all younger female users.

86.7% of all older male users succeeded in accomplishing the search task successfully versus 78.6% of all older female users.

Older users with a higher level of education were slightly more successful than less educated older users: 84.2% versus 80.0%.

Older people using the internet daily were more successful than older people who did not: 88.9% versus 72.7%.

4.4.2 Efficiency (Total Time Spent on Search Task)

In order to analyse how efficient the users were in accomplishing their search task, only those users who succeeded in accomplishing this successfully within 5 minutes were taken into account.

User Group	Average search time (in seconds)	Median (seconds)	N
All users	86	63	51
All older users	111	76	24
All younger users	64	45	27
All female users	66	58	23
All male users	102	72	28
All younger female users	63	36	12
All younger male users	64	48	15
All older female users	69	60	11
All older male users	146	77	13
All older users with higher education	112	76	16
All older users with no higher education	108	74	8
All older users using internet daily	99	86	16
All older user not using internet daily	135	77	8

Table 4: Efficiency (Source: Loos & Mante-Meijer (2009, p.62))

The most striking results to emerge were:
Overall, users took an average of 86 seconds to accomplish their search task.

Younger users were almost twice as fast as older users, averaging 64 seconds as against 111 seconds. The eye-tracking studies conducted by Tullis (2007) and Houtepen (2007) showed a similar result (see section 2).

Female users were almost twice as fast as male users, averaging 66 versus 102 seconds. However, it should be borne in mind that the older men in the study took an extremely long time to complete the task. Older male users and older users who did not make daily use of the internet were the slowest (respectively averaging 146 seconds and 135 seconds). Older female users, younger male users and younger female users were the fastest (respectively averaging 69 seconds, 64 seconds and 63 seconds).

Older users making daily use of the internet were faster than those who did not, averaging 99 versus 135 seconds.

4.4.3 User Satisfaction (Usability Rated by 1-10)

User Group	Average note	Median note	N
All users	6.8	7	58
All older users	7.2	7	29
All younger users	6.4	7	29
All female users	6.7	7	28
All male users	6.9	7	30
All younger female users	6.1	6.5	14
All younger male users	6.7	7	15
All older female users	7.2	7	14
All older male users	7.1	8	15
All older users with higher education	7	7	19
All older users without higher education	7.,5	8	10
All older users using internet daily	7.4	7.3	18
All older user not using internet daily	6.9	7	11

Table 5: User satisfaction (Source: Loos & Mante-Meijer (2009, p.63))

The most striking results to emerge were:

- Overall, users took an average of 86 seconds to accomplish their search task.
- Younger users were almost twice as fast as older users, averaging 64 seconds as against 111 seconds. The eye-tracking studies conducted by Tullis (2007) and Houtepen (2007) showed a similar result (see section 2).
- Female users were almost twice as fast as male users, averaging 66 versus 102 seconds. However, it should be borne in mind that the older men in the study took an extremely long time to complete the task. Older male users and older users who did not make daily use of the internet were the slowest (respectively averaging 146 seconds and 135 seconds). Older female users, younger male users and younger female users were the fastest (respectively averaging 69 seconds, 64 seconds and 63 seconds).

- Older users making daily use of the internet were faster than those who did not, averaging 99 versus 135 seconds.

5 Conclusions

5.1 The Role of Age

The results of this explorative case-study demonstrate in the first place that, to some extent, age has an impact on the way older and younger users accomplish a search task on a website, especially where efficiency and user satisfaction are concerned.

Younger users were found to be almost twice as fast as older users, averaging 64 versus 111 seconds. The eye-tracking studies conducted by Tullis (2007) and Houtepen (2007) showed a similar result (see section 2).

Another finding was that older users tended to be more satisfied with the website than the younger ones. This could be explained by the fact that the website belongs to an organization for older people and that the information is related to life events which are more relevant for older than for younger users.

Yet this explorative case-study also showed that age is not the only variable explaining differences in navigation behaviour. Gender, educational background and frequency of internet use all also play a role.

5.2 The Role of Gender, Educational Background and Frequency of Internet Use

5.2.1 Gender

- All younger male users succeeded in accomplishing the search task successfully versus 85.7% of all younger female users.
- Female users were almost twice as fast as male users, averaging 66 versus 102 seconds due to the fact that older men are extremely slow.
- Average user satisfaction of younger female users was much lower than average user satisfaction of younger male users: 6.1 versus 6.7.
- Younger male users needed fewer saccades to reach the webpage containing the relevant information than younger female users.

5.2.2 Educational Background

Older users with a higher educational level tended to be less satisfied with the website than those with a lower level of education: 7 versus 7.5. Otherwise, educational background did not play an important role.

5.2.3 Frequency of Internet Use

The heatmaps clearly show that daily internet use has more impact on our navigation patterns than age does.

- Older people using the internet daily were more successful than older people who did not: 88.9% versus 72.7%.
- Older users making daily use of the internet were faster than those who did not, averaging 99 versus 135 seconds.
- Older users who use the internet daily tended to be much more satisfied with the website than those who did not: 7.4 versus 6.9.

5.3 Final Conclusion

Finally, we can conclude that, although differences in navigation behaviour are to some extent age-related, differences are also evident within the group of older people (what Dannefer (1988) refers to as *intra-age variability*) depending on gender, educational background and frequency of internet use. In this case study, the black-and-white distinction between Prensky's *digital natives* and *digital immigrants* was absent. Instead, what emerged was far more a digital spectrum rather than a digital divide (Lenhart & Horrigan, 2003). If future empirical research confirms the findings of this explorative eye-tracking study, the implication for website designers (who often belong to a younger generation) might be that they should take into account diversity between and within generations by *designing for dynamic diversity* (Gregor, Newell & Zajicek, 2002), '(…) the premise of which is that older people are much more diverse in terms of life experience and levels of capability and disability than their younger counterparts (…)' (Chisnell & Redish, 2004, p. 48).

Bibliography

Andrew, A., 2008. Web Accessibility for Older Users: A Literature Review. W3C working draft 14 May 2008. Available at: http://www.w3.org/TR/wai-age-literature [Accessed May 2009].

Balazs, A., 2004. Marketing to Older Adults. In: J. Nussbaum & J. Coupland, eds. 2004. *Handbook of Communication and Aging*. London: Lawrence Earlbaum Associates.

Bouma, H. 2000. Document and interface design for older citizens', In: P. Westendorp, C. Jansen & R. Punselie eds. 2000. *Interface design & document design*. Amsterdam: Rodopi.

Chisnell, D. & Redish, J., 2004. Designing Web Sites for Older Adults: A review of recent research. Available at: http://www.aarp.org/ olderwiserwired [Accessed May 2009].

Chisnell, D. & Redish, J., 2005. Designing Web Sites for Older Adults: Expert Review of Usability for Older Adults at 50 Web Sites. Available at: http://www.aarp.org/olderwiserwired [Accessed October 2009].

Council of the European Union and the Commission of the European Communities, 2000. Action Plan eEurope – An Information Society For All. Brussels

Dannefer, D., 1988. What's in a Name? An Account of the Neglect of Variability in the Study of Aging. In: J.E. Birren & V.L. Bengtson, eds., 1988. *Emergent theories of aging*. New York: Springer.

Frøkjaer, E., Herzum, M. & Hornbaek, K., 2000. Measuring Usability: are Effectiveness, Efficiency, and Satisfaction Correlated? Proceedings of the SIGCHI Conference on Human Factors in Computing Systems. Den Haag.

Goldberg, J.H. & Wichansky, A.M., 2003. Eye Tracking in Usability Evaluation. In: J. Hyönä, R. Radach & H. Deubel, eds., 2003. *The Mind's Eye: Cognitive and Applied Aspects of Eye Movements Research*. Amsterdam: Elsevier.

Gregor, P., Newell, A.F. & Zajicek, M., 2002. Designing for Dynamic Diversity - interfaces for older people. In: *ASSETS 2002*, pp.151-156.

Houtepen, L., 2007. *Op zoek naar inFormatie. Onderzoek naar het vinden en Beoordelen van Informatie op de Websites van de vijf Grootste Zorgverzekeraars*. [Unpublished Master thesis] Utrecht, Utrecht University / Utrecht School of Governance.

Johnson, R. & Kent, S., 2007. Designing Universal Access: Web Application for the Elderly and Disabled, *Cogn Tech Work* 9.

Kalmijn, M. et al., 2006. Kwaliteit van leven: De Dynamiek van Levenslopen. Concept-Notitie voor NWO. Den Haag: NWO.

Lenhart, A. & Horrigan, J.B., 2003. Re-visualizing the Digital Divide as a Digital Spectrum. *IT & Society*, 5, pp. 23-39.

Loos, E.F., 2009. User-centred websites: The (Ir)relevance of Age. Include 2009 proceedings (International conference on Inclusive Design), Royal College of Art, London, 5-8 April 2009.

Loos, E.F. & Mante-Meijer, E.A., 2009. *Navigatie van Ouderen en Jongeren in beeld. Explorerend Onderzoek naar de rol van Leeftijd voor het Informatiezoekgedrag van Website-Gebruikers.* Den Haag: Lemma.

Nussbaum, J. & Coupland, J. eds., 2004. *Handbook of Communication and Aging.* London: Lawrence Earlbaum Associates.

Pernice, K. and Nielsen, J., 2002. *Web usability for senior citizens. Design Guidelines Based on Usability Studies with People Age 65 and Older.* Fremont: Nielsen Norman Group.

Prensky, M., 2001. Digital Natives, Digital Immigrants. *On the Horizon*, 9(5), pp.1-6.

Sap, J.C.M., 2004. *Leeftijdsloze Levensloop.* Utrecht: Expertisecentrum LEEFtijd.

Stephanidis, C., 2007. *Universal Access in Human Computer Interaction. Coping with Diversity. Reihe: Lecture Notes in Computer Science (LNCS).* New York: Springer.

Thomése, F., 2001. Het levensloopperspectief in theorie en onderzoek. Een sociaal-gerontologische blik. In: Y. Quispel & L. Christ, eds 2001. *Ouder worden: Een kwestie van leeftijd?* Utrecht. LBL.

Tullis, T., 2007. Older Adults and the Web: Lessons Learned from Eye-Tracking. In: C. Stephanidis, ed. 2007. *Universal access in human computer interaction. Coping with diversity.* Reihe: Lecture Notes in Computer Science (LNCS). New York: Springer.

Wichansky, A.M., 2000. Usability Testing in 2000 and Beyond. *Ergonomics* 43(7), pp. 998-1006.

Chiara Carini, Ivana Pais

Business Social Networks: an 'Age Levelling' Service?

1 Business Social Network Services: Definition and Theoretical Assumptions

Over the past two decades there has been an enormous increase in interest regarding the role of social networks in shaping economic phenomena such as labour markets, international trade, migration and entrepreneurship etc. Most of the literature is about the economic effects of social networks; less is known about the factors that generate, sustain and reproduce these networks (Smith-Doerr & Powell, 2005).

In recent years this field of study has been concerned with the emergence and unexpected success of several organizations which focus on building, making visible and maintaining business social networks: business social network services (BSNSs).

In order to better understand the boundaries of this concept it is important to define each component:

- a network is something that is composed of nodes (vertices) and ties (edges);
- in a social network, the nodes are represented by actors and the ties represent the relations between them;
- business social networks are professional – not personal – social networks
- business social network services are intentional organizations; they can be in presence (business club) or online (business social network sites).

In line with boyd and Ellison (2008), we define them as business social 'network' services and not business social 'networking' services: while networking (relationship initiation, often between strangers) is one of the objectives of these services, it is not the primary practice for many of them, nor is it what differentiates them from other services. What makes business social network services unique is that they enable users to articulate and make visible their business social network.

When analysing BSNSs one can refer to two main branches of literature: the one related to economic sociology and the role of social capital within business social networks, and the branch related to social media studies and social network sites.

1.1 The Perspective of Economic Sociology

Mark Granovetter's study (1973) on the "strength of weak ties" in linking labour supply and demand drew attention of social scientists to the role of personal relations. The strength of a tie is "(...) a (probably linear) combination of the amount of time, the emotional intensity, the intimacy (mutual confiding), and the reciprocal services which characterize the tie" (Granovetter, 1973, p.1361). Granovetter claimed that weak ties are more useful in occupational mobility because they provide people with access to information and resources beyond those available in their own social circle.

After consideration of the limited role traditionally assigned to relational networks, Granovetter went on to advance his thesis on the embeddedness of economic transactions within social networks and created the so called "New Economic Sociology".

Though this term was not used explicitly, the policy manifesto of the New Economic Sociology (Granovetter, 1985) made a fundamental contribution to the debate on social capital, started at the time by Pierre Bourdieu (1980) and achieved thorough theoretical formulation in James Coleman's work (1990). The meeting of these two currents of analysis gave rise to an internationally extended independent research programme focusing on two essential issues: how social networks influence behaviours of networkers and how these individuals can use networks to pursue their specific aims.

Scholars in the field of network research have developed conceptual instruments that help us understand how differences in network structures facilitate different forms of social capital. Among these, Ronald Burt's (1992) concept of structural holes describes how individuals could benefit more from a few contacts to unconnected groups than from multiple connections to people who all know each other.

Over the last decade, this line of research has strongly developed in the direction of the concept of "social resources" (Lin, 2001). Lin sees social networks as the foundation of social capital: he defines social capital as resources embedded in the social structure, which are accessed/mobilised through intentional actions. His theory is based on the assumption that one invests in social relations in expectation of profit, via the resources that flow within. The profitability of social capital is linked to four reasons: it eases the flow of information, makes it possible to exert influence on decision makers, certifies social credentials and reinforces individual identity.

In this literature, though with different emphasis according to the disciplinary domain and research tradition, the social capital mobilised in social networks is mostly considered a by-product of activities initiated for other purposes, aimed at objectives differing from those originally pursued. Coleman speaks of 'appropriable social organisations' to indicate the possibility of orienting a network of relations towards new aims compared to those originally pursued and links this possibility to the so called 'multiplexity', a concept used by anthropologists to show the multiple nature of relations. Coleman himself does not rule out the existence of explicit and intentional attempts to build social capital for economic purposes and, thus, he makes the distinction between appropriable social organisations and 'intentional organisations'; however, his attention is mostly focused on the former.

The new feature drawing social scientists' attention in recent years, compared to traditional studies on the role of social capital in the market (and, in particular, the labour market) rather lies in the emerging (and strengthening, where they already exist) of business social network services as organisations intentionally aimed at creating – as well as mobilising – social capital directly pursuing business purposes.

With the end of the industrial society and the crisis of the Fordist-Keynesian paradigm, came a change in the characteristics that 'knowledge workers' are required to have and, therefore, the criteria for selecting and evaluating employees, but a similar change has also taken place regarding the factors leading to success for freelance workers and small-scale entrepreneurs. The spread of new forms of organisations and of a "new culture of capitalism" (Boltanski & Chiapello, 1999), and also the progressive erosion of collective social capital (Putnam, 1993), favour and reward the mobilisation of individual social capital for economic purposes. Such context conditions lead to a greater awareness of the economic role of social resources, produce a strong demand for spaces (both physical and virtual) aimed at establishing and maintaining social relations and, finally, create new spaces of sociality expressly oriented towards building individual social capital, which is subsequently useful for enhancing workers' employability and company competitiveness.

1.2 The Perspective of Social Media

Since 2001 almost 200 social networking websites focusing on both personal and professional relations have been launched. These sites have attracted the attention of academic and industry researchers who are drawn to their affordability and reach (boyd & Ellison, 2007). This is a very recent field of study and, at the moment, there are no specific publications regarding social network sites focused on business. However, in order to better analyse this subject, one can refer to the main results obtained by researchers who have analysed the Internet and general social network sites.

Internet is a social space where older relationships can be maintained and new ones can be formed (Baym, 2006). While some researchers argue that computer-mediated communication technology, such as the Internet, is too limited for the creation of meaningful relationships (Nie & Erbring, 2002), others claim that the Internet actually makes people more social (Walther, 1996). The very definition of the word *social* is often a key element to these contrasting statements (Olsen, 2008).

Many people use the Internet with the intention of asserting and/or exploring their identity. Given the Internet's multiple character and ability to adapt to its user's needs, a person will often be able to find an application that suits him/her (Baym, 2006). For example, they may form new relationships by joining online groups in search of a sense of belonging, information, empathy or social status.

The Internet is no longer a separate world for the millions of people that routinely go online (Vaage, 2007). Rather than isolating users in a virtual world, the Internet extends communities in the real world, and connects people through individualised, flexible social networks rather than fixed grounded groups (Wellman, 2004). The main difference between online forums or communities and social network sites is that the former are primarily organised around people rather than interests.

Boyd and Ellison (2007) define social network sites as web-based services that allow individuals to (1) construct a public or semi-public profile within a bounded system, (2) articulate a list of other users with whom they share a connection, and (3) view and traverse their list of connections and those made by others within the system. The nature and nomenclature of these connections may vary from site to site.

The rapid growth of online social networks is creating a new social environment able to change our linguistic codes, semantics and social syntax (Mazzoli, 2009). People are "becoming the media" and learning to communicate with "networked publics" (Ito, 2008), and through these practices they are increasing their self-reflexivity (Boccia Artieri, 2009). User-generated contents are – at the same time – more public and more private (Giglietto, 2009). Simply the public display of connection can be seen as an identity signal that helps people explore the networked social world (Donath & boyd, 2004).

The main characteristic of online social networks is that they make the "invisible links" that tie people together in daily life both visible and perceivable (Boccia Artieri, 2009). For this reason, a social network service does not necessarily diminish the number of chains between people but, by giving them more visibility, makes it easier to choose the 'right' paths. Social networking services may provide a bridge between online and offline social relationships, and they are particularly effective in relation to the maintenance of weak ties, such as former classmates or colleagues (boyd & Ellison, 2007).

As for face-to-face relations, not all the personal contacts on a social network site are equally important: an average Facebook user has about 120 friends but he actually talks to about four to six people on a regular basis (Cameron, 2009); with instant messaging people have an average of approximately 100 contacts on their list, but they chat with less than five and 80% of phone calls are actually made to the same four people. But within online social networks even weak links become stronger: a connection is valuable because it is a potential means of transfer; social networks increase the possibility of results, not the certainty.

2 Does Age Influence Access and Participation in Business Social Networks?

In order to analyse how social capital is intentionally created, maintained and mobilised in business social networks services, we carried out an exploratory analysis of Milan IN, a non-profit association set up in 2005 to allow members of LinkedIn living in Milan to physically meet up with each other. Milan IN now registers more than 5,300 members. LinkedIn is the largest business-oriented social network site, launched in May 2003 in Mountain View, California, with more than 50 million users worldwide, spanning more than 200 countries (11 million users in Europe, 800 users in Italy).

We were thus able to analyse both online (LinkedIn) and face-to-face (Milan IN) business social networks, and to highlight any differences perceived by people who use them both.

The following research methods were used:

- participant observation (by registering with LinkedIn and Milan IN website and attending meetings between September 2008 and November 2009)
- an online questionnaire, available between May 15th and 22nd 2009, 1,674 Milan IN members out of a total of 5,387 took part (31%)
- semi-structured interviews with the two youngest and the two oldest Milan IN members
- entering a LinkedIn forum, asking the question "Do you think that your age is affecting your participation in LinkedIn?"

One of the most interesting results of our research concerns the relationship between the age of Milan IN members and their approach to both online and face-to-face business social networks. This relationship can be analyzed from three different points of view.

2.1 Does Age Influence the Decision to Join a Business Social Network?

Milan IN has only one requirement for new members; they already have to be registered with LinkedIn, which implies a sort of self-selection by people who speak English and are technologically literate. Apart from that, anyone who subscribes to the rules and distinctive values of the organisation is accepted.

> "Milan IN was founded [...] with the aim of gathering people coming from different professional areas, who believe in networking as a vital business support, have strong motivations, love their job and consider their skills to develop profitable business as a form of human creativity." (www.milanin.com)

The age bracket of Milan IN members is quite broad, ranging from 22 to 84, but 61% of its members were born between 1966 and 1978. The average age is 39. This finding is not surprising: Milan IN is not generally used either by young people, due to its focus on business, or by seniors who are less familiar with technology.

The most interesting finding concerns the reason behind the registration of under-30s and over-60s : they registered with LinkedIn (and then with Milan IN) to avoid age-related prejudice; they think that, unlike face-to-face business, within business social network sites, experience matters more than age.

> "With these tools age differences do not exist. When you meet someone in person, age is something that affects the relationship, with LinkedIn the first impression is not based on age and even when you meet off line, age continues to be irrelevant"
> (Milan IN, man, temporary manager, 71 years old)

> "I dress in a way that makes me appear older, I have grown a beard for that reason. Because otherwise you don't work. Via internet you don't need it"
> (Milan IN, man, entrepreneur, 22 years old)

> "Participants here tend to be more experienced, which is not linked to age, gender or ethnicity" (LinkedIn Q&A, man, consultant)

> "I cannot say that I've even thought about it, so I'd say no. Social media is an equalizer in that respect; what matters is what you know, how you share, and the time you put into it"
> (LinkedIn Q&A, man, General Manager at Web Industries)

Age is also strictly related to the way people discover business social networks. This is particularly evident when observing LinkedIn.

47% of interviewees had found LinkedIn by word of mouth. This communication tool is particularly effective among young adults: 29% of those born before 1953 joined LinkedIn following a recommendation by friends, colleagues or others; this percentage increases to 42% for people born between 1953 and

1965 and to 49% for those born between 1972 and 1978 and it reaches the maximum percentage (54%) for people under 31.

34% of members received an invitation to join LinkedIn. People in their forties and fifties mostly use this communication tool. The data grouped by age reveals that the percentage of people who received an invitation is highest for those born before 1953 (44%) and this percentage went down as the age of the interviewees decreased: 26% of people born after 1978 received invitations.

Only 15% of people report having discovered LinkedIn in the general and specialised media. The data shows no significant differences between age groups.

Even for Milan IN, word of mouth is the most important means of communication (35% of those interviewed), especially for young members, with only 16% reporting an invitation to join and 11% finding out through the media.

2.2 Does Age Influence How Users Participate?

As far as other personal and professional characteristics are concerned, Milan IN members are predominantly:

- men (79% compared to 21% women), even if women in Italy represent the highest rate of participation to LinkedIn compared to other European countries (36% compared to 35% in Spain, 34% in France, 32% in the UK, 31% in the Netherlands and 29% in Germany);
- highly educated: 80% have a degree and 39% of them have a postgraduate education; in Italy, only 11.6% of the economically active population has a degree, compared to the European average of 23.2% (Eurostat 2005);
- 57% are employed, 18% entrepreneurs, 17% self-employed, 6,5% unemployed and only 1% students and 0,5% retired;
- in ICT (27%) and consultancy (19%)

Milan IN members are people with a high level of education, who mainly operate in areas where good knowledge of the web is fundamental and who use its opportunities to improve their position at work.

Milan IN is, by now, quite a homogeneous group and this may explain why there are no significant differences in how members participate.

The data collected from the online survey supports this hypothesis and confirms that not even age affects how people use LinkedIn and Milan IN.

The majority of the Milan IN members use LinkedIn regularly:

- 22% of interviewees have daily access to their profile on LinkedIn, with rates ranging from 21% of those born between 1953 and 1965 to 24% of those born between 1979 and 1985.
- 27% access it 2 or 3 times a week, with rates ranging from 20% of those born between 1979 and 1985 to 29% of those born between 1953 and 1965;
- 21% check their profile once a week, with rates ranging between 19% of those born between 1953 and 1965 and 25% of those born after 1978;
- 14% access it 2 or 3 times a month, with rates ranging from 9% of those born before 1953 to 15% of those born between 1953 and 1965.
- there are no differences based on age even between interviewees who use LinkedIn rarely, once a month or less (16%): the percentages range from 13% of those born between 1966 and 1971 to 20% of those born before 1953.

Looking at the change in how people use LinkedIn:

- 39% of people access with the same frequency; these percentages range from 38% of people born between 1972 and 1978 to 43% of those born between 1953 and 1965;
- 32% of interviewees access their profile more than in the past, with rates from 31% of those born between 1953 and 1965 to 36% of those born before 1953;
- 29% of interviewees use LinkedIn less than in the past, range from 23% of those born before 1953 to 30% of those born after 1978.

Milan IN members have, on average, 224 contacts in LinkedIn, and most of them (77%) have an "open strategy" (Olsen, 2008): they connect with people they do not know personally (on average, 30% of the total number of contacts).

People use the Milan IN site less than LinkedIn and only 27% of respondents participate in the meetings organised by the association. Of all members who participate in events:

- 89% do it occasionally;
- 7% often participate;
- only 4% take part in every meeting.

The meetings organised by the association are held in Milan. Low participation in the meetings depends largely on the fact that only half of the respondents live (47%) or work (55%) in the city of Milan. Work and family commitments make it difficult to attend meetings; as proof of this, 68% of members who never participate would if they had the chance. Participation in meetings is linked to two factors in particular: the agenda of the meeting (44%) and whether time is available (36%).

2.3 Does Age Influence Outcome?

"We consider Milan IN as a community of people, from the professional and corporate world, interested in expanding beyond the traditional boundaries of their company [...] the network of contacts in the work and business world to create more job opportunities for themselves and for others, to create new business opportunities between members, to broaden the knowledge of the professional world different from its own." (www.Milanin.com)

LinkedIn and Milan IN are used for 4 main purposes:

- to create new contacts (networking): 70% of LinkedIn users and 58% of Milan IN users;
- to maintain existing contacts: 65% of LinkedIn users and only 18% of Milan IN users;
- to be updated: 42% of LinkedIn users and 39% of Milan IN users;
- to reconnect with people no longer in contact in real life: 40% of LinkedIn users and only 11% of Milan IN users.

Age does not influence the importance of these four purposes: for each age group the main aim is to find new business contacts. In particular, people born before 1953 consider it very important to use LinkedIn to look for new contacts (79% compared to approximately 60% of the other age groups) and less important to re-establish lost contacts or to maintain existing contacts (40% compared to 65%).

Creating new contacts is the main goal for the association members, but it is the hardest to achieve through LinkedIn:

- only 34% created new contacts (compared with 70% who considered it to be an objective);
- 58% maintained existing contacts (compared with the expected 65%);
- 38% kept updated (compared with the expected 42%);
- but 59% of members reconnected with people they used to know (compared with the expected 40%);
- 16% of respondents say they did not achieve anything from using LinkedIn: 6% of them say they it was a waste of time, while 94% are very confident about future results. Age-related differences are minimal: the number of people who report no achieved results vary between 14% of those born after 1978 and 20% of those born before 1953.

The number of respondents who obtained results with Milan IN is lower than that for LinkedIn. Even in this case age does not affect the results:

- 18% of members found new contacts with Milan IN (compared with the expected 58%);
- 13% kept updated (compared with the expected 39%);
- 8% maintained existing contacts (compared with the expected 18%);
- 4% re-established lost contacts (compared with the expected 11%).

The number of people who did not obtain results by joining Milan IN is higher than for LinkedIn: 48%. Again the majority of this group (82%) is confident about obtaining results in the future.

Considering the results obtained and the opportunities created, Milan IN members rate the two business social networks as satisfactory. On a scale from 1 to 10 people give an average an average score of 6.3 for LinkedIn. In particular the lowest opinion is expressed by people over 56 (average score 6), while people under 32 are the most satisfied, with an average score of 6.5. There is a relationship between frequency of use and satisfaction: people who access LinkedIn daily or weekly are more satisfied with the results than occasional users.

5.5 is the average rate for Milan IN: average ratings are between 5.4 expressed by people born between 1953 and 1966 and 5.9 of people born before 1953. Users who participate in meetings are more satisfied with the site (6.1). Average ratings range from 5.9 expressed by people born between 1966 and 1971 and to 6.7 for people born before 1953. Users who have never participated in meetings give a lower score (5.1 on average): average ratings are between 4.9 reported by people born between 1953 and 1965, and 5.5 reported by people born after 1978.

These results confirm our theoretical assumption: business social networks are not evaluated by the results obtained, but by the potential members perceive. Even if they have not obtained the results they expected, they are confident about the future.

3 Conclusions

Looking back at our research questions, it is now possible to give some answers.

- Does age influence the decision to join a business social network? Yes.
- Does age influence how people participate? No.
- Does age influence the results obtained? No.

Business social networks are mainly used by people born between 1966 and 1978. The younger and older age groups use them to avoid age-related prejudices. They want the opportunity to prove their expertise, and not to be judged by their age.

Age discrimination is probably the least recognised and understood of the various social prejudices which affect the life and, in particular, the work of people in Europe (Midwinter, 2005). Both younger and older people may be affected by stereotypes. For example, younger people are often assumed to lack maturity; older people are often assumed to lack cognitive capacity, physical strength and endurance, flexibility, motivation, and the ability to promote and absorb new ideas.

Because of these assumptions, mature jobseekers are being excluded from the recruitment process because they are perceived as being 'too old' to do certain jobs. If this waste of skills, experience and competence among mature professionals is counter-productive today, it will become an even greater problem in the future as Europe's demographic profile changes (www.mature-project.eu). The opposite problem affects young people, in particular in Italy, where the social mechanisms of generational turnover are slower than in other European countries (Ambrosi & Rosina, 2009).

The Eurobarometer survey "Discrimination in the European Union 2009" shows that 54% of Italians interviewed think that age discrimination is widespread in their country. Compared with results from 2008, there is a notable increase (+8 percentage points); the economic crisis seems to be behind this shift in opinion. In the previous 12 months, 6% of those interviewed felt personally discriminated against or harassed on the basis of their age and 8% witnessed someone else being discriminated. Only 38% of Italian respondents (55% EU average) think that enough is being done to increase diversity in their workplace as far as age is concerned.

In 2000, the European Union adopted two very far-reaching laws to prohibit discrimination in the workplace based on racial or ethnic origin, religion or belief, disability, age or sexual orientation.

A business social network can be seen as a way to combat (or, at least, avoid) age discrimination, as an "age levelling" service: once people access it, age does not matter, how people participate and the results obtained are equally distributed.

Future research could further test this hypothesis; in particular, it would be interesting to compare members of business social networks with their colleagues and do a social network analysis (Sna) of connection and recommendations in order to see if there is any generational difference in the business social network structure.

Acknowledgements

We would like to thank: the Board of Directors of Milan IN, and in particular Pier Carlo Pozzati and Silvia Lenich; Concept srl who gave us the opportunity to use MyK to manage our online survey; and all those who took part in the survey. Without their collaboration, this research would not have been possible.

Bibliography

Ambrosi E. & Rosina A., 2009. *Non è un paese per giovani. L'anomalia italiana: una generazione senza voce.* Venezia: Marsilio Editori.

Baym, N.K., 2006. Interpersonal Life Online. In: L.A. Lievrouw & S. Livingstone, eds. 2006. *The Handbook of New Media.* London, California and New Delhi: Sage Publications Ltd.

Boccia Artieri, G., 2009. SuperNetwork: quando le vite sono connesse. In: Mazzoli, L., 2009. ed. Network effect. *Quando la rete diventa pop.* Torino: Codice edizioni, pp.21-40.

Burt, R., 1992. *Structural Holes.* Cambridge: Harvard University Press.

Coleman, J., 1990. *Foundations of Social theory.* Cambridge: Harvard University Press.

Donath, J., boyd, d., 2004. Public displays of connection. *BT Technology Journal*, 22(4), pp.71-82.

Giglietto, F., 2009. Io, i miei amici e il mondo: uno studio comparativo su Facebook e Badoo in Italia. In: Mazzoli, L., 2009. ed. *Network effect. Quando la rete diventa pop.* Torino: Codice edizioni, pp.41-56.

Granovetter, M., 1973. The Strength of Weak Ties. *American Journal of Sociology.* 78(6), pp.1360-1380.

Haythornthwaite, C., 2000. Online personal networks. Size, Composition and Media Use among Distance Learners. *New Media and Society*, 2(2), pp.195-226.

Ito, M., 2008. Introduction. In: V. Kazys, ed. 2008. *Networked publics*, Boston: MIT Press.

Lin, N., 2001. *Social Capital: A Theory of Social Structure and Action.* New York: Cambridge University Press.

Mazzoli, L., 2009. Introduzione. In: Mazzoli, L., 2009. ed. *Network effect. Quando la rete diventa pop.* Torino: Codice edizioni, pp.3-20.

Marlow, C., The Structural Determinants of Media Contagion, Phd Thesis. Available at: http://cameronmarlow.com/papers/phd-thesis [Accessed 8 December 2009].

Midwinter, E., 2005. How many people are there in the third age?. *Ageing & Society*, 25 (1), pp.9-18.

Olsen, L.E., 2008. Professional networking online. A qualitative study of LinkedIn use in Norway, Thesis. Available at: University of Bergen Open Research Archive https://bora.uib.no/handle/1956/2935 [Accessed 8 December 2009].

Smith-Doerr, L. & W., Powell, 2005. Networks and Economic Life. In: N. Smelser & R. Swedberg, 2005. *The Handbook of Economic Sociology*. Princeton: Princeton University Press.

Wellman, B., 1988. Structural analysis: From method and metaphor to theory and substance. In: B. Wellman & S. D. Berkowitz, ed. 1988. *Social Structures: A Network Approach*. Cambridge, UK: Cambridge University Press, pp.19-91.

Wellman, B., 2004. Connecting Community: On- and Off-line. *Context*, 3(4), pp.22-28.

Tanja Oblak Črnič

The Generational Gap and Diverse Roles of Computer Technology: The Case of Slovenian Households

1 Introduction

The main characteristic of communication technologies is their wide presence in our homes. According to the Statistical Office of the Republic of Slovenia, in 2007 two thirds of Slovenian households (66%) owned a personal computer, 55% of them having access to the Internet. Ten years before, a personal computer could be found in only one third of Slovenian households (Oblak, 2008, pp.163-164). Nowadays computer technologies along with other, well-established, media present an important factor in the construction of our privacy. At the same time, computer technologies are subject to constant domestication. While these technologies are a source of novelties in family relations, contributing to the changes of the home environment, the response to them is defined by the family's relationships, their inner structures of power, gender and also age.

The process of "domestication" is related to the capability of a social group – in our case the family or the household – to accept various technological artifacts into their own culture to such an extent that these technologies become "invisible" within the daily life of the group (Silverstone & Hirsch, 1992). Starting with the assumption that the relationship of the individual towards modern technical appliances bears different weights, this article is concentrated on one of the rare ethnographic studies of the uses of computer technologies in Slovenian families.

In the late nineties, when the "computer technoculture" began to spread into Slovenian homes, the personal computer had to fight its way into the rooms that had been dominated by the old media, especially radio, video and TV. However, research focused on generational differences in the usage of new media in the Slovenian context has been quite rare. More often the research has been focused on media usage by the Slovenian population in general (Luthar et al., 2001) or to specific topics related to media use in the younger generation (Dolničar & Nadoh, 2004). In order to fill the missing link this paper tries to reveal the crucial differences in the domestication of personal computers between the parents, usually the initiators of the purchase of a computer, and their children, who have eventually taken over the new technology and become its chief caretakers.

A big difference exists in the understanding of what the computer is when it comes to parents and their children. While most parents understand the computer as a dividing line between home and work, secondary-school students often perceive it as a tool which helps them with home obligations and school activities. While young people perceive computer technology as a bridge between various structures of everyday life, their parents often consider that same technology as a strictly dividing factor. The domestication of computer technologies is therefore "not only about appropriating objects to the self, but is about how we make ourselves at home in our everyday environments, how we make them habitable and comfortable, and use objects to manage the social world" (Lalley, 2002:2).

2 Slovenian Media Culture in the Private Sphere

The material culture of homes in modern society is - also in relation to the presented data - intensified, which can be seen in the arrangement of rooms, in the growing presence of sophisticated technologies and in the redefinition of standards, criteria and hierarchy of values (Putnam, 1992, p. 195). The activities within a household do not depend merely on the family structure, but also on the integration of family members into wider social structures outside their home. New technologies, especially those connecting the family with the outer world, give new characteristics and meanings to the private sphere. Consequently, differentiated ways of use and different media habits emerge.

2.1 Diffusion of Personal Computers and the Internet in Slovenian Households

As has been established before (Oblak, 2008, p.163), according to Slovene Public Opinion the increase in the number of personal computers used in the private sphere in Slovenia began in 1994 and has gone up from 15% (April 1994) to 50% (November 2001) of households using computers. Table 1 shows the increase of the number of households possessing a personal computer according to age, where a difference between the youngest and the middle generation is obvious.

	Jan. 91	Nov. 93	Apr. 94	Nov. 95	Jun. 96	Dec. 97	Nov. 98	Nov. 99	Dec. 2000	Nov. 2001
<30	16%	16%	17%	28%	30%	41%	44%	57%	62%	69%
30-46	15%	18%	19%	26%	32%	39%	39%	57%	55%	64%
46-60	9%	8%	17%	22%	23%	29%	37%	39%	46%	50%
>60	3%	5%	5%	5%	7%	9%	12%	16%	16%	14%
TOTAL	12%	13%	15%	21%	24%	32%	34%	45%	46%	49%

Table 1: The percentage of Slovenian households with a PC, divided by age, from 1991 to 2001 (Source: Slovene Public Opinion).

According to Statistical Yearbooks (see Table 2), in 1996 only 3% of Slovenian households had access to the Internet. In 1999 the number increased to 15% and in 2007 58% of Slovenian households had access to the Internet. In 2007, the personal computer was already present in two thirds (66%) of Slovenian households.

% of households with	96	97	98	99	0	1	2	3	4	5	6	7
personal computer	24	32	35	42	46	47	58	55	58	61	65	66
Internet access	3	8	9	15	21	24	37	40	47	48	54	58
mode of access												
Modem					82	71	55	49	45	29	15	
ISDN					15	17	12	19	16	13	11	
ADSL					...	2	7	15	24	39	50	
Cable					3	7	14	7	18	27	34	
Wireless					...	...	2	38	47	44	43	

Table 2: Slovenian households with personal computers and Internet access from 1996 to 2007 (Source: Statistical Yearbooks)

The numbers above strongly suggest a rise of Internet access and acceptance of personal computers within Slovenian households over the last decade. However, the beginnings of computer culture were still rather modest – in 1996 only 3% of households had Internet access, which rose only to 15 % in 1999. Nowadays, the old modem mode of Internet access is vanishing and is being substituted by much more sophisticated and faster Internet connections.

Table 3 shows how Internet access in Slovenian households has been divided over time by gender and age structure; the early adopters were men younger than 30 years, who were caught by younger women only after 2002. In all these years women above 50 years belong to a group with the smallest access to the Internet within the private sphere.

	1996		1997		1999		2000		2001		2002		2003		2005		2006	
	M	W	M	W	M	W	M	W	M	W	M	W	M	W	M	W	M	W
under 30	4	2	28	22	31	18	37	26	42	38	52	42	58	57	70	66	91	85
30-50	3	4	19	25	23	20	41	42	30	33	43	37	52	52	53	59	68	72
above 50	2	3	14	12	12	8	14	7	20	14	25	15	25	20	35	25	49	38

Table 3: *Percentage of men and women with Internet access divided by age from 1996 to 2006 (Source: Slovene Public Opinion)*

An important dividing factor between offline and online households in all these years remains the level of education. The divide between the group with secondary or higher school education on the one hand and with primary school only on the other is getting even bigger; as the data in Table 4 shows, in 2005 only one third of less educated men or one quarter of less educated women had Internet access, while the percentage of higher educated owners is almost 40 % higher for both genders.

	1996		1997		1999		2000		2001		2002		2003		2005		2006	
	M	W	M	W	M	W	M	W	M	W	M	W	M	W	M	W	M	W
OŠ+PŠ	5	5	9	10	10	6	13	9	15	12	18	11	26	22	31	25	52	37
SŠ+VŠ	4	6	30	29	35	24	39	31	41	38	58	48	62	58	70	66	81	80

Table 4: *Percentage of men and women with Internet access divided by education from 1996 to 2006 (Source: Slovene Public Opinion)*

2.2 Media Consumption among Slovenian Youth

An additional view on what is happening behind the family front door is provided by looking at media usage in the young generation. What do they watch? How much of their spare time is devoted to media usage? In what way do they expose themselves to different – new and old - media? What meanings do young people give to different media in their everyday life? Can they imagine their daily life without TV or mobile phones? A set of similar question was given in some research done in Slovenia in 2001 on 519 youngsters from 14 to 19 years old; a study which included 243 pupils from primary and 276 students from secondary school[1]. In order to give a narrow but interesting glimpse into the media habits of the Slovenian younger generation, some of the most relevant data that helped to understand the general picture of media usage in Slovenia are presented in the following section.

According to this research, in 2001 the most important communication technology for young people was the mobile telephone, followed by the personal computer and the Internet. An interesting result shows that the primary importance of the mobile phone was not determined by any demographic statistics (age, gender, parental education, etc.), like for instance the Internet. The meaning of Internet and personal computers in everyday life was much higher for those youngsters, whose fathers have higher education (Dolničar & Nadoh, 2004, p. 33). Additionally, the data showed that young people, mostly female, who live in urban places miss personal computers and the Internet more often than the others.

The access to *old* media in private homes was almost unquestionable; TV (99%), radio (98%), books (97%) and telephones (96%) have been strongly adopted in Slovenian households of the included respondents; only access to cable/satellite TV and to CD players has been statistically determined by the type of local community (urban vs. rural) and age (Dolničar and Nadoh 2004: 39); those who live in bigger towns and who are older have easier access to them. However, as was mentioned already before, the access to *new* media is mostly determined by the respondents' age or gender (Table 5) and most strongly by the level of parental education (Table 6).

---

[1] The survey included 7 Slovenian primary schools and 7 secondary schools from different regions within different age classes: from 12-13 years, 14-15 years, 16-17 years and 18-19. The questionnaires were answered in class, supervised by teachers, from January to February 2001. The majority of respondents are female (60% vs. 40 % male), they live in rural areas (47%), 28% in smaller towns and 25% in bigger cities.

		computer or video games	personal computer	Internet	mobile phone
All		80	81	47	82
Gender	male	85*	84	56***	85
	female	77*	79	41***	80
Age	12-13	77	72*	47	70**
	14-15	82	84*	47	79**
	16-17	79	80*	44	86**
	18-19	84	87*	50	89**
Local community	bigger town	85	85	58*	87*
	smaller town	80	81	46*	83*
	village	78	79	42*	77*

Table 5: Access to new media among the Slovenian younger generation divided by age, gender and type of local community (Source: Dolničar and Nadoh 2001)

		computer or video games	personal computer	Internet	mobile phone
Primary school	mother	59***	60***	9***	71
	father	59***	60***	17***	76
Secondary school	mother	80***	80***	44***	81
	father	80***	82***	42***	79
High school	mother	87***	90***	62***	84
	father	92***	90***	62***	84
Master or doctoral degree	mother	83***	88***	70***	87
	father	87***	93***	72***	93

Table 6: Access to new media among the Slovenian young generation divided by parental education and their gender (Source: Dolničar & Nadoh, 2001)

Parental education not only affects the access to new media, but is also an important "regulator" of the quantity of media exposure. The young people with higher educated parents spend much more time on the Internet and behind the computer than the others. Additionally, young Internet users usually do better in school, as has been noticed also in other countries, such as the UK (see Livingstone & Bovill; 1999).

But to thoroughly understand media usage it is not enough if we analyze the quantity of exposure (or access) to different media and technology. More often it becomes much more relevant if we understand how people consume media and what kind of meanings they give to different media in their everyday life. The trend of individualization of media consumption that has been revealed in many other studies (Livingstone, 1998; Ule et al., 2000) provides in this sense an important theoretical context for the interpretation of the so called "media bedroom culture". An important question in this view is related to the private context of media use, namely, to the preferences for "collective" or "individual" ways of media consumption. Here, at least in the Slovenian case, the difference between "TV supported media" vs. "PC supported media" has been recognised; young people would prefer to use the Internet and PC alone, without the presence of either of their parents, while they would prefer to watch TV shows with their closest friends (Dolničar & Nadoh, 2004, p.90).

What could we summarise from the data presented so far? Slovenian households have been materially strongly equipped with different communication technologies and media, although this depends mostly on socially determined factors like age, education and gender. While for old media - like TV, radio and also the newest personal technology, the mobile phone - these differences are rather small or statistically not relevant, the opposite seems to be the case for PC supported media. In order to give a detailed study focused on the role of computer technologies in Slovenian homes we present another case study that will help to reveal very different meanings and understandings of such technologies between two generations: parents and their secondary-school children.

3 Analysing Family Roles of Computer Technology

As various foreign research shows (Bakardjieva, 2005; Facer et al., 2003; Holloway & Valentine, 2000; Lalley, 2002), the introduction of the computer as a communication technology into the private sphere of the home, which was still ruled by the dyad of radio and television at the beginning of the nineties, is neither coincidental nor should it be taken for granted. The introduction of personal computers is often considered to be a result of the "dominant rationality of productivity and efficiency defining computer technology that had brought the machine into the homes" (Bakardjieva, 2005, p. 93). However, this is not the only user discourse about the role of computers present in various interpretations about the role of the computer in the family. On the contrary, such studies confirm that the spreading of different communication technologies in society is a complex process supported by various motivations and accompanied by unexpected turns, which greatly influence family life, the ways families spend their time together and even the patterns of life in the home environment. Talking about their personal computer as a new "family member" is often accompanied by claims about its indisputable negative influence on family relations.

One approach that primarily derives from cultural and media studies, provides an understanding of various forms of adopting new technologies and their adaptation to everyday-life environments, especially those of homes (Bakardjieva, 2005, p.24). The so-called model of "domestication of technologies", developed in the early nineties by a well-known Birmingham research group, stressed the active role and the meaning of context in the reception of the media among various groups of audience. The project focused on the patterns and contexts of "technology consumption" as material artefacts and domesticated media. The approach presupposed that the production of technologies, which is the result of many environments, contexts, decisions and interests of various institutions, does not end with the "dispersion" of these technologies at home, but continues with the consumption of them; consumption involves the inclusion of an object in the personal and social identity of the consumer (Gell in Hirsch, 1992, p. 209), which consequently means that the objects are eternal in nature, as they live through the form of the social relationships that they help to generate (Gell in Hirsch 1992, p.209). New technologies are becoming a part of everyday routine, but at the same time they are changing it. They enter everyday life in a way that everyday routines become constructed around the objects, which turn out to be unnoticed through the process of domestication (Bausinger, 1984, p.346). Therefore, an object is not merely missed because of its functionality, but because the absence of this object disrupts daily routine[2].

In order to understand how family members are able to develop personalised relationships of ownership, various types of "domesticated computer cultures", conditioned with the technological artefacts of everyday life, are presented with the help of comparable studies about the usage of computers within the private sphere:

- The first research uses the project "Screenplay" (Facer et al. 2003), which ran from 1998 to 2000 in Great Britain. With the help of this project the scientists analysed various uses of computer technologies among young people in their homes as well as in school environments[3].

---

2 This was already proved in the nineties by studying various cases of radio (Moores, 1988), television (Silverstone, 1994), satellite TV (Moores, 1996) and personal computer (Haddon, 1992) consumption.
3 The results used in this paper are the ones acquired in the study about the usage of computers at home in a group of eighteen youngsters, and with the help of interviews with children and their parents, and through the observation of computer usage. For details see Facer et al. (2003).

- A few years later another research study was done by Bakardjieva with the intention of showing various uses of computer technologies and the Internet within families in Canada[4].
- An independent Slovenian research study, which focuses on the connection between the parents' vocabulary about the domestication of computer technologies and the interpretations of these technologies that are adopted by the younger family members[5]. All the studies mentioned above are based on the model of domestication of modern technologies. All the analyses are done in a qualitative way mainly through interviews and observations.

In order to show the typical categories of computer roles in Slovenian families we compare "harmonious speech", which interprets the computer as something positive, welcome, useful or even necessary, with a more "conflicting speech", which regards the computer as the cause for internal misunderstandings and potential conflicts.

3.1 Computer Roles through the Parents' Vocabulary

How do parents respond to the computer as a technological object? When do they buy it and what do they intend to use it for? How do they (re)shape its meaning within their family life?

---

4   In her study, Bakardjieva (2005) concentrated on the way a "home" Internet user is formed, how the "Internet room" is formed in a home, how the interactions between family members and the family values change, and how the user manages his or her "virtual blending". She acquired the results through interviews with 23 volunteers of both sexes, of different ages and of different education. For details see Bakardjieva (2005, pp.76-91).

5   The population included in the research was a group of nine families, where one parent and one secondary school member were interviewed; in sum, eighteen people were interviewed. They belonged to the rural, suburban and urban population and to both sexes. The structured interviews, which were partly done by students on the course Communication and New Technologies, were done according to a pre-planned scheme in November and December 2006. The identities of the interviewees are not revealed in the article. The interviews included questions about the introduction of computer technology into the home and the reasons for it, about the way of objectifying these objects, and about the actual forms of use and changes in the field of interpersonal relations.

3.1.1 Inevitable Member of the Family

One of the crucial conclusions about the relationship between the computer as an artefact of technology and the household as a private sphere is that the computer is not simply an "exclusive domestic artefact" but enters the sphere of the home primarily as "a technology associated with the workplace" (Facer et al., 2003, p.32). For this type of family it is characteristic that the computer is interpreted as an "extension of the parents' work". Here is an example of a mother, who explains the reason for the purchase of a computer: "When my husband started his own trade, we had to buy a computer for him to be able to save information, enter data and print." (mother of two children, 37). Another example shows the necessities of work for a family father: "Yes, I needed the computer mainly because of my work. I have my own firm and this requires a great deal of work at home. I think this was the main reason for the purchase." (father of two children, 52).

Some other studies show that most families regard the computer as a necessary companion in the lives of their children, either as a way of spending their free time or as a possibility for their future work. Here, the computer was seen as the child's machine, as a key symbolic representation of the aspirations of parents for their children's future welfare (Facer et al., 2003, p. 34). In these families, where the computer is considered mainly as an "opportunity for the children", the computer was located in a room either belonging to a child or used exclusively by the children.

Slovenian parents, although to a smaller extent than elsewhere, also prove that the computer is considered as a "useful" help or a necessity intended mainly for the children's education: "Yes, we bought the computer mainly for the children, because they needed it for their school work. (...) I think my son wanted to have a computer more than any of us." (father of three children, 50).

The link between the first two types of understanding the role of a computer in families crystallises itself in the prototype, where the computer is, above all, regarded as an "inevitable investment for the future". This discourse was the most common one in the interviews with Slovenian parents. Let us have a look at the example of a father, who proudly explains that, by buying a computer, he "saved" the family from being behind the times: "… I brought the computer into our house, mainly because it was a novelty then and I did not want my family to be deprived of novelties. The children, especially the older son, constantly played computer games. Later I started using the computer for my work." (father of two children, 53). Another view, which is even more marked by the "inevitability" of buying a computer, is one of a father of three children:

"The reason for the purchase was a very simple one: we needed a computer. It was popular, many people had it already, it was supposed to bring us advantages. So, we bought one. It was necessary, just like a TV." (father of three children, 45)

This last example points to a necessity enforced by society, determined by the pressures of "popularity", and the fact that a particular technology is already present elsewhere. Above all, this necessity is enforced by the ascribed meaning about the useful advantages, which one cannot miss. This way, the investment for the future is conditioned by the discourse about a functional, progressive and absolutely necessary technology.

3.1.2 The Conflicting Faces of a Computer: from the Intruder to the Comforter?

Besides the above mentioned "harmonic images" of the computer's role, which are the proof of relatively unproblematic interventions of the computer in the home environment, interviews with parents also showed the less pleasant and even problematic sides of the new computer culture.

One type of family can be described as a not completely domesticated space, because these families do not, according to Silverstone's model, include the computer into their own life frames. This "absence of domestication" is usually limited to a single member of the family. Therefore the image the computer has is as an "intruder". The computer still remains a necessary 'fact of life', brought reluctantly into the home (Facer et al., 2003, p.39), since the absence of a computer would represent an obstacle for the children, who cannot compete with others without it. Let us take the following example: "The computer represented great fun and an adrenalin rush along with the desire to learn new things. My husband enjoyed the computer, as well, and at the same time it made his work easier. Personally, I have always considered the computer as an intruder and I am still not very fond of it." (mother of two children, 50)

The Slovenian study shows another type of conflicting interpretation: the computer being regarded as a "trouble-maker". The parents blame it for the family conflicts and consequently, it is supposed to lead to a diminished communication among the family members.

> "I've lost my temper several times, because there is nobody I can talk to. Everyone sits behind their 'box' and surfs the Internet, etc. Sometimes it seems as if we are complete strangers. Even when we quarrel, we never talk about it afterwards, because everyone pretends to be busy behind their computer and supposedly does not have time to talk. Sometimes I am so angry about all this technological progress!" (mother of two daughters, 47)

That the computer is something "bad" is shown in the following example, as well: "I think that the computer is a backup. Everybody has become lazy. Even the personal relations are not what they used to be. Nobody moves around anymore. I strongly object to the fact that the computer could ever become the heart of home. I think it is more of an enemy to good and healthy relations." (mother of two children, 47)

The third type of conflicting vocabulary of parents about computers has an indirect nature. Often children find comfort in front of the computer screen, which their parents find unacceptable. Here are the words of a father, who finds the computer to be an "inappropriate comforter":

> "… I have noticed several times that, after a quarrel, my children find comfort in computer games, which I do not approve of. I would rather see them crying on their beds than sitting in front of the computer screen, because this is certainly not the right solution." (father of three children, 45)

3.2 Ambivalent Roles of Computers for the Youngsters

The parents' images of computers, which are expressed through the memories of "the family biography" of the home computer are numerous and are mutually exclusive. They are important because they act as models for the parents' introduction of the "appropriate" use of computers to their children, who can either accept these encoded meanings or reject them by establishing their own uses. Therefore, the explanation of various computer usages in the homes also demands that various youngsters' views of what a computer is are taken into account; these views are conditioned by the school experience of young people, by the peer culture and by the media culture in general.

Compared to the parents' representation of the computer, one of the most common, even dominant representations of what a computer means for young people, is that of "transitional nature"; due to the range of uses and experiences that youngsters in secondary schools have nowadays, their statements show the transition from "the computer as a toy" to "the computer as a multi-tool for every occasion". Two examples speak in favour of such finding:

> "At the beginning, the computer was just a game for me. I played the same computer game day after day and kept comparing my current results to those from the previous days. However, today the computer means everything to me. I never run out of ideas of what to do with it …" (youngest son, 17).

Or another example:

"At the beginning the computer was a toy. However, today this toy has acquired a new meaning. It has become a multi-tool for everything from searching for cribs and seminar essays, to free-time activities and chatting with friends. I have even flirted with girls in various chat-rooms. I also use the computer for searching music and movies, which I then listen to or watch on the computer, as well. Everything is possible if you have a computer." (middle son, 16)

This transition is conditioned by the difference in what the computer offered in its pre-Internet phase and what it can offer now. Judging by the interviews we could say that the Internet gave an entirely new meaning to the computer. This is especially true in the case of younger family members, who keep stressing that the Internet means everything to them. The computer is therefore not merely a study tool, but serves as a social and communication space, where friendships are sustained, new partners are found, deals are made and free-time activities are formed.

However, it is not necessary that this transition of the computer from being a toy to being a multi-tool for every occasion is always considered as a positive change. Here is a statement of a secondary-school student, which points to the "limitations" of computers today: "When I got my computer, it was just a new toy, which I did not really need and it served mostly for playing computer games. Now it represents a nuisance, because I have to write different school papers and assignments. Computer games are of no interest to me any more." (younger daughter, 18). This ambivalent role of the computer is determined by the additional "serious" identity that a computer, connected to the Internet, can have.

The third group of youngsters and their understanding of computers is very protective about this personal tool, because they do not relate the technology to the "bad" family situations as their parents do. What is more, some young people even claim that the computer "should not be blamed" for these bad family relations. This implies the desire for neutralization of everyday phenomena that have to do with technology. Let us look at the following example: "Maybe we do not talk as much as we used to, since everyone is in his or her own corner, doing things for work or for school. (…) We are so preoccupied with ourselves that we do not find the time for deep discussions any more, but I think that the reason for this is not the computer. It is just how things are today; people, rushing and not having any time for each other." (youngest daughter, 17) Even though young people, after thinking about it, admit that growing up also means entering a period of less communication with the family, they do not necessarily ascribe this to technology as their parents do. Young people search for reasons outside their homes, or compare the possible effects of computer technology as a way of spending free time to the time spent in front of the television screen, which has also "atomised" family members. This is also the case in the following example, which does not recognise any changes in interpersonal communication:

> "Everything is the same! The relations have not changed, although some people might claim otherwise (...) Usually, the parents would say that. Although parents always think that they do not spend enough time with us anyway (...) but this is not true. Young people spend more time in front of the computer screen than in front of the TV. This is how you kill time and you do not talk so much. But in the times before the computer and the Internet, there was the television, right?" (son, 18)

Young people understand the reasons for the transformed ways of communication as a consequence of the changed structure of spending time, which is marked with serious school and work-related tasks.

4. Conclusion

Summing up all these studies we can conclude that, in the private sphere of today, the computer has many roles, which are connected with the parents' expectations and aspirations as well as with their fear of what is going on behind the computer screen. According to interviews, young people perceive these generation differences well. They do not ascribe the changes in spending time and ways of managing potential family conflicts to the technology itself. On the contrary, a new structure of time consumption burdened with serious school and work-related activities is given as the main reason for the changes in communication. Although their everyday lives seem to be led by the computer screen, young people tend to find reasons for the lack of communication and social life in the family also outside the computer screen.

Although computer technology firstly served as a mere "machine" or "toy", used mainly for business or spare time activities, that is not the case anymore. Today the computer has taken on a whole new level of existence. It is not considered only as "a good investment for the future" and an aid in the education of children. The computer is also thought of as an "intruder" and a "destroyer" of interpersonal relations on the one hand and a "comforter" in family disputes on the other.

The Internet has given the computer a new "communication identity" through its Web platform. The forms of social activities have widened and the Web is now an intrinsic part of young people's lives; here they search for information, acquire materials for their school projects, complete their school tasks, communicate with each other, search for new friends, listen to music and watch films. Their lives have become mirror images of Web environments. Contrary to their children, the parents understand the computer as a relatively separated part of their lives, although they use the computer to organise their lives too. Still, their history and the way the computer was brought into their homes, remain a dominant sign of their momentary misunderstanding of what is going on in the youngsters' rooms and the youngsters' patterns of using the Internet.

Bibliography

Bakardjieva, M., 2005. *Internet Society: The Internet in everyday life*. London: Sage.

Bausinger, H., 1984. Media, Technology and Daily life. Media. *Culture and Society*, (10)6, pp. 343-351.

Dolničar, V. & Jana, N., 2004. *Medijske Navade med Slovenskimi Mladostniki*. Ljubljana: FDV.

Facer, K., Furlong, J., Furlong, R. & Sutherland, R., 2003. *ScreenPlay: Children and Computing in the home*. London: RoutledgeFalmer.

Green, L., 2000. *Communication, Technology and Society*. London: Sage.

Haddon, L., 1992. Explaining ICT Consumption: the Case of Home Computer. In: R. Silverstone & E. Hirsch ed. *Consuming Technologies: Media and Information in Domestic Space*. London: Routledge.

Haddon, L. & Silverstone R., 2000. Information and Communication technologies and Everyday Life: Individual and Social Dimensions. In: K. Ducatel, J. Webster & W. Herrmann ed. *The Information Society in Europe*, New York: Rowman&Littlefield.

Hirsch, E., 1992. The Long Term and Short Term of Domestic Consumption: and Ethnographic Case Study. In: V R. Silverstone & E. Hirsch ed. *Consuming Technologies: Media and Information in Domestic Space*. London: Routledge.

Holloway, L. S. & Gill V., 2003. *Cyberkids: Children in the Information Age*. London: RoutledgeFalmer.

Kopytoff, I., 1986. The Cultural Biography of Things: Commodification as Process. In: A. Appadurai. *The Social Life of things: Commodities in Cultural Perspective*. Cambridge: Cambridge University press.

Lalley, E., 2002. *At Home with Computers*. Oxford: Berg.

Livingstone, S. 1998. New Media, new Audiences? *New Media&Society*, 1, pp.59-66.

Livingstone, S. and Bovill, S., 1999. *Young People, New Media*. London: Sage.

Livingstone, S., 2008. Taking Risky Opportunities in Youthful Content Creation: Teenagers' Use of Social Networking Sites for Intimacy, Privacy and Self-expression. *New Media&Society*, 10(3), pp. 393-411.

Luthar, B., 2007. Mobilni Telefon in Pospešena Kultura. Javnost/The Public, suplement, vol XIV, pp.5-18.

Miller, D. & Don S., 2000. *The Internet: an Etnographic Approach.* Oxford: Berg.

Moores, S., 1988. The Box of the Dresse: Memories of Early Radio and Everyday Life. *Media, Culture&Society*, 10(1), pp.23-40.

Moores, S., 1996. *Interpreting Audience: the Etnography of Media Consumption.* London: Sage.

Oblak Črnič, T., 2007. Med Simbolnimi Pomeni in Realnimi Praksami Mobilnega Vsakdanjika. In: V. Vehovar ed. *Mobilne Refleksije.* Ljubljana: FDV

Oblak Črnič, T., 2008: O Začetkih Interneta na Slovenskem. In: M. Pušnik ed.*Prispevki k Zgodovini Medijev na Slovenskem*, Javnost/The public, Suplement.

Putnam, T., 1992. Regimes of Closure: the Representation of Cultural Process in Domestic Consumpton. In: R. Silverstone & E. Hirsch ed. *Consuming Technologies: Media and Information in Domestic Space.* London: Routledge.

Rogers, E. & Larsen J., 1984. *Sillicon Valley Feaver: Growth of High Technology Culture.* New York: Basic books.

Schiffman, L., Bednall D., Watson J. & Kanauk L., 1997. *Cosumer Behaviours.* Sydney: Prentice Hall.

Silverstone, R., 1994. *Televison and Everyday Life.* London: Routledge.

Silverstone, R. & Hirsch E., 1992. *Consuming Technologies: Media and Information in Domestic Space.* London: Routledge.

Strathern, M., 1992. The Mirror of Technology. In: R. Silverstone & E. Hirsch ed. *Consuming Technologies: Media and Information in Domestic Space.* London: Routledge.

Ule, M. et al., 2000. *Socialna Ranljivost Mladih.* Urad RS za mladino. Ljubljana.

Authors

Aroldi, Piermarco (PhD), is Associate Professor of Sociology of Cultural and Communicative Processes at the Università Cattolica of Milan and Piacenza, where he teaches Media and Social Processes. He is vice-director of Osservatorio sulla Comunicazione. His research interests are the temporal dimension of media, TV consumption in everyday life and from a generational perspective, Media and children. Among his publications are: *I tempi della Tv. La Televisione tra offerta e consumo* (Carocci, Roma 2007) all of which deal with time/space and the media; *Le età della Tv. Indagine su quattro generazioni di spettatori italiani* (co-edited with Fausto Colombo, Vita e Pensiero 2003) and *Successi culturali e pubblici generazionali*, Milano, Link, 2007 (co-edited with F. Colombo), both of which are about generational identity in media consumption; *Il gioco delle regole. Tv e tutela dei minori in sei paesi europei* (Vita e Pensiero, 2003), which deals with the co-regulatory system relating to minors' protection.

Bayraktutan-Sütcü, Günseli is a Ph.D. candidate at Ankara University Graduate School of Social Sciences. She currently teaches as a full time lecturer at Baskent University, Faculty of Communication. Her master thesis was on knowledge industries and communication networks (2004). She is writing her dissertation on the effect of new media on the political participation among academicians. Her recent publications are *Kültür Endüstrisi Olarak Dijital Oyun* (2008) (co-authored), co-edited a book on digital game culture in Turkey, *Dijital Oyun Rehberi* (2009), *Observatorio* (OBS*) 3:1 (2009) Available at : http://www.obercom.pt/ojs/index.php/obs/article/view/249/247, and a book chapter "Practising Identity in the Digital Game World: The Turkish Tribes' Community Practices in 'Silkroad Online'" (co-authored), in *Digital Technologies of the Self* (Ed. by Yasmine Abbas and Fred Dervin). UK, New Castle: Cambridge Scholars Publishing, pp.61-85. Her research interests are new media and political communication studies.

Binark, Mutlu is Professor at the Department of Radio-Television and Cinema, Faculty of Communication, Baskent University. She teaches media sociology, intercultural communication, and new media culture. She is currently working on digital game culture and media pedagogy. Her recent publications are *Internet, Toplum, Kültür* (2005) (co-edited), *Eleştirel Medya Okuryazarlığı* (2007) (co-authored), *Yeni Medya Çalışmaları* (2007), *Kültür Endüstrisi Olarak Dijital Oyu*n (2008) (co-authored), *Toplumsal Paylaşım Ağı Facebook* (2009) (co-authored). *Dijital Oyun Rehberi* (co-edited) (2009), a book chapter (co-authored) "A Critical Evaluation of Media Literacy in Turkey and Suggestions for

Developing Critical Media Literacy for Citizenship and Democratic Social Transformation" *Information and Media Literacy.* (Ed. M. Leaning), Information Science (2009), an article in *Observatorio* (OBS*) 3:1 (2009), and recently a book chapter "Practising Identity in the Digital Game World: The Turkish Tribes' Community Practices in 'Silkroad Online'" (co-authored), in *Digital Technologies of the Self* (Eds. Yasmine Abbas and Fred Dervin), Cambridge Scholars Publishing.

Boccia Artieri, Giovanni is Professor of Sociology of New Media in the Faculty of Sociology at the University Carlo Bo of Urbino (Italy) and vice-director of LaRiCa (Research Laboratoy in Advanced Communication). He is board member of AIS (Italian Association of Sociology). Editorial co-director of "MediaCultura" and "Mediologie" book series, member of the editorial board of the journal "Sociologia della Comunicazione". His research interests deal with the relationship between media, society and identity, languages and expressive forms of modernity at large. Among his latest publications: SuperNetwork: quando le vite sono connesse, in L. Mazzoli (ed), Network Effect. Quando la rete diventa pop (Torino 2009); Share This! Le culture partecipative nei media. Un'introduzione a Henry Jenkins (Milano 2008), Forme e linguaggi dei videomondi: dai videogame a SecondLife (Napoli 2008), I media-mondo. Forme e linguaggi dell'esperienza contemporanea (Roma 2004).

Carini, Chiara collaborates in the Centre for Studies and Research Data, Methods and Systems at the Department of Quantitative Methods, University of Brescia (Italy). She got her Degree in Theory and methods for information management at the Faculty of Economics, University of Brescia. Her research deals with the study, collection and statistical analysis in the field Economic and social development.

Centorrino, Marco is a researcher in Sociology of cultural and communication processes at the Faculty of Humanities, University of Messina (Italy), where he teaches Sociology of Communication. Recently he has worked on issues related mainly to mass communication, in its traditional and innovative forms and publishing: *Bulli, pupe e videofonini* (Bonanno, 2008), *La rivoluzione satellitare* (Franco Angeli, 2006). His research has moved along three main areas: media representation of political communication; use of new technologies, in particular electronic entertainment industry; studies of major media, according to an en-coding-decoding scheme, focusing on mass phenomena associated with so-called popular culture.

Colombo, Fausto is Professor of Media Theories and Media and Politics at the Faculty of Political Sciences, Università Cattolica del Sacro Cuore, Milan, and he has been Director of Osservatorio sulla Comunicazione since 1994. He planned and carried out research for several bodies and he is a member of many national and international editorial boards. He is member of the Scientific Committee of Triennale di Milano, a foundation dealing with design, fashion and audio-visual communication. One of his present interests is technological innovation and generational identity in media consumption. He has published several books and essays, the most recent book being *Le età della Tv*, Vita e Pensiero, Milano, 2002 (edited with P. Aroldi); *Tv and Interactivity in Europe. Mythologies, theoretical perspectives, real experiences*, Vita e Pensiero, Milano 2004 (ed); *Atlante della comunicazione*, Hoepli, Milano 2005 (ed); *Digitasing TV. Theoretical Issues and Comparative Studies across Europe*, Vita e Pensiero, Milano 2006 (edited with N.Vittadini); *Successi culturali e pubblici generazionali*, Milano, Link, 2007 (edited with P. Aroldi*)*; *La digitalizzazione dei media*, Carocci, Roma 2007 (ed).

Contarello, Alberta is Professor of Social Psychology at the University of Padova, Italy. Her current interests concern the social construction of knowledge - with particular regard to the new technologies - social psychology and qualitative research, social psychology and literary texts. Among her publications on the theme of ICTs: Contarello, A. (2003) 'Body to body: co-presence in communication', in *Mediating the Human Body. Technology, Communication and Fashion*, eds L. Fortunati, J.E. Katz & R. Riccini, pp. 123-131. Contarello, A. & Fortunati, L. (2006) 'ICTs and the human body: a social representation approach', in *New Technologies in Global Societies*, eds P. Law, L. Fortunati & S.Yang, pp. 51-74.

Corsten, Michael is Professor of Sociology at University of Hildesheim. He studied sociology, economics, social history and philosophy at the universities of Marburg and of Bielefeld. PhD in sociology at the University of Marburg, habilitation (second PhD) at the Free University of Berlin. He worked as researcher at the Institute of Sociology in Marburg, at the Max-Planck-Institute for Human Development in Berlin and at the University of Jena. Research Interests: Biographies and Cultures of Generations, pragmatist and discursive social theory. Selected Publications: The Times of Generations (in: Time & Society, 1999); Biographical revisions and the Coherence of a Generation (in: *Mayall/Zeiher*, 2003, *Childhood in Generational Perspectives*, London).

Črnič, Tanja Oblak is an Associate Professor in Department for Media and Communication Studies and researcher at the Research Centre for Social Communication at the University of Ljubljana (Slovenia). Her research is focused on e-democracy, changes of political communication on the web, interactivity of online media and social dimensions of internet use in everyday life.

Among her publications: *Dialogue and representation: Communication in the electronic public sphere. Javnost* (Ljubl.), 2002; *The lack of interactivity and hypertextuality in online media. Gazette* (Leiden). 2005; *Slovenian online campaigning during the 2004 European parliament election.* Kluver, R. et al. *The Internet and national elections: a comparative study of web campaigning*, (London; New York), Routledge, 2007.

Dolničar, Vesna, PhD, is a researcher at the Centre for Methodology and Informatics, Faculty of Social Sciences, University of Ljubljana. She is pedagogically engaged in several courses (e.g. Survey Methodology, Internet in Everyday Life, Virtual Research Environment) and has been actively involved in many (inter)national research projects related to the field of understanding the information society and the specific needs and motives of the potentially excluded groups, particularly the elderly. She is currently a national correspondent for the 6th FP project SOPRANO (Service-oriented Programmable Smart Environments for Older Europeans), a management committee member of the COST Action 298: Participation in the Broadband Society and a principal researcher of a postdoctoral project titled 'The relationship between e-inclusion and social inclusion'. She (co)authored several scientific papers, a scientific monograph and five chapters in the monographs of publishers Wiley, Greenwood and Peter Lang.

Fortunati, Leopoldina is Professor of Sociology of Communication at the University of Udine. She has conducted several research in the field of gender studies, cultural processes and communication and information technologies. She is the author of many books and articles and serves as referee for several journals. She is the co-chair with Richard Ling of the International Association "The Society for the Social Study of Mobile Communication" (SSSMC) which intends to facilitate the international advancement of cross-disciplinary mobile communication studies. She organised many international workshops and conferences. Her works have been published in eleven languages. Among her recent publications on the theme of ICTs: Fortunati, L. (2006) 'Mobile communication and its payment', *Asian Communication Research*, Vol. 3, No.1, pp. 21-47; L. Fortunati (2006), 'User design and the democratization of the mobile phone, *First Monday*, special issue number 7, September 2006.

Hardey, Mariann is a researcher at the University of York. She got a MA with distinction in Social Research in 2005 and a Ph.D. degree in 2008 with a thesis titled Seriously Social: Making Connections in the Information Age. This research charts the rise of digital social networks and associated media in the lives of young people - commonly known as the Generation-Y. Her latest publications include Hardey, M. (2009) Key Trends: The Social Context of

Online Market Research: An Introduction to the Sociability of Social Media, *WARC Journal of Market Research*, 51(4) July.; Hardey, M. (2009) Productive Consumers, report to Innovative Media for a Digital Economy (IMDE) cluster, University of Oxford. Hardey, M. (2008) The formation of rules for digital interactions. *Information, Communication & Society*, vol. 11, no. 8, pp. 1043-1045. And she continues to write the popular properfacebooketiquette.com blog.

Kortti, Jukka, adjunct Professor (docent), is a social and media historian from University of Helsinki, Finland. His doctoral thesis (2003) was about the Finnish post-war modernization studied through television advertising. His has also published a comprehensive social historical study (2007) about 50 years of Finnish television. At the moment, he is writing the 100 years history of the student magazine *Ylioppilaslehti*, which is not "any student paper", but a significant Finnish cultural and political institution. His latest international publications include for instance: Jukka Kortti and Tuuli Anna Mähönen: 'Reminiscing television: Media Ethnography, Oral History and Finnish Third Generation Media History' *European Journal of Communication* 2009, vol 24(1): 49–67.

Loos, Eugène, as a linguist, is a Professor of Old and New Media in an Ageing Society at the University of Amsterdam and a Senior Lecturer at Utrecht University in the Netherlands. He conducts research on the ways in which older people looking for information use old and new media. He was a member of the COST Action 298 where he coordinated the research on "New media in the hands of different generations: Designing an inclusive society". He published several (inter)national research papers, articles and books in this field, and edited with Leslie Haddon and Enid Mante-Meijer the book *The Social Dynamics of Information and Communication Technology*, published by Ashgate in 2008.

Mortara, Ariela is assistant professor at IULM University of Milan.
Her research topics are sociology of consumption, corporate communication, advertising and branding. Since 1995 she teaches at IULM University and since 2005 at the University of Trento. Her most recent publications are: "Dalla vita nello schermo alla vita sullo schermo: i social network", Canestrari P., Romeo A (2010) (eds.), *Dall'uomo all'avatar e ritorno. Realtà e dimensioni emergenti*, Verona, QuiEdit, Vol. 1. "Toward the construction of a museum brand: the case of the Museo Nazionale Della Scienza e della Tecnologia Leonardo Da Vinci of Milan", proceedings of IX International Congress Marketing Trends, Venice, 21-23 January 2010,."The rise of ethical fashion: a sociological perspective" and (with Anna Maria Bagnasco), "Product placement: a new tool to manage for the art sector?", proceedings of the 9th Annual Conference of the European Sociological Association (ESA 2009), Lisbon 2 -5 September 2009.

Müller Sonja is research consultant at empirica. She holds a degree in geography, urban development, and economic sociology from Bonn University. Main fields of her research activity are ICT applications and their usage patterns amongst older people and people with special needs and the way ICTs impact on societal participation and inclusion of disadvantaged population groups. Since joining empirica in 2005 she has been involved in a number of projects focusing on the development and evaluation of technical systems in the area of Independent Living/Ambient Assisted Living and policy analysis in the area of enclusion. Sonja Müller has authored and co-authored national and international publications.

Pais, Ivana is assistant professor of Economic Sociology at the University of Brescia (Italy), Faculty of Law. She teaches Sociology in the Information Society, Sociology of work and Human Resources Management. Her reseach deals with new jobs and new ways of working in the information society. Among her publications: *Acrobati nella rete. I lavoratori di internet tra euforia e disillusione*, Franco Angeli, Milano 2003.

Romaioli, Diego is Psychologist, Psychotherapist and PhD. He's also Professor at the University of Padua (Italy) and an associate member of the Institute of Psychology and Psychotherapy of Padua and the International Institute Taos. He got a Ph.D. in Psychological Sciences at the Department of General Psychology in Padua. He deals with psychosocial research and its importance in the context of clinical psychology and human organizations. He is an expert on advanced models of change in psychotherapy. Among his publications: Romaioli D., Faccio E., Salvini A. (2008). On Acting Against one's Best Judgement: a Social-Constructionist Interpretation for the Akrasia Problem. *Journal For The Theory Of Social Behaviour*, 38 (2), p. 179-192.

Santi, Marco, is a psychologist specialised in cognitive ergonomics, user experience and user research methods. He followed the project called SOPRANO (Service-oriented Programmable Smart Environments for Older Europeans). His main interest in the SOPRANO project was to create methods involving older users in all stages of the process developing assistive technology.

Sarrica, Mauro is currently employed as a post-doctoral researcher at the Department of Applied Psychology - Padua University (Italy).He is a social psychologist interested in social construction of knowledge more specifically, social representations. His research focuses primarily on social representations of peace, war and conflict. During the last four years he has been working on projects related to social construction of Environmental Conflicts. He is doing research in other areas as well, including active citizenship, representations of ICTs, the impact of the internet in newsrooms, positive aging. Among his latest publications: Fortunati, L. et al., (2009). The influence of the Internet on European journalism. *Journal of Computer Mediated Communication*, 14(4), 928-963; Contarello, A. & Sarrica, M. (2007) 'ICTs, social thinking and subjective well-being. The internet and its representations in everyday life'. *Computers in Human Behaviors*, vol. 23, pp. 1016-1032.

Siibak, Andra received her degree in Media and Communication (2009) from the University of Tartu, Estonia, where she is a research fellow in media studies at the Institute of Journalism and Communication. Her research interests include online content creation practices of young people, visual and textual self presentation in social networking websites and gender identity constructions in virtual environments. Currently she is a member of the research group of cultural communication studies in the Centre of Excellence in Cultural Theory and involved in the research projects "Children and Young People in the Emerging Information and Consumer Society", financed by the Estonian Science Foundation and "Construction and normalization of gender online among young people in Estonia and Sweden" financed by The Foundation for Baltic and East European Studies.. Her articles have appeared in international journals including *Journal of Computer Mediated Communication, Cyberpsychology, Journal of Children and Media*.

Treleani, Matteo is a PhD fellow at the Institut national de l'audiovisuel in agreement with Paris Diderot University. His research, directed by Marc Vernet and Bruno Bachimont, investigates how the change of context affects the meaning of archival videos. He thus looks for new ways to recontextualise heritage documents. He got his degree in Semiotics at Bologna University (Italy) in 2007. He's also a journalist and part of the selection committee of the Festival International Jean Rouch (Comité du film Ethnographique) since 2008.

Vellar, Agnese is a PhD candidate in Communication Science at the University of Torino (Italy). Her research interests are Tv and music fandom from a transnational perspective and the role of media in the construction of individual and collective identities. She worked at *Technology, Research and Trends Department - Telecom Italia* conducting qualitative research on social media and designing interface for digital television. She is conducting workshops on media ethnography at the University of Torino.

Index

action,
    political, 38, 63
    social, 104, 105, 201

adult, 10, 13, 23, 33, 41, 64, 121, 122, 125, 127, 129, 150, 160, 163, 165, 166, 167, 168, 187, 189, 193, 202, 205, 209, 211, 213, 216, 217, 222, 225, 226, 227, 228, 229, 231, 245, 261, 262, 280

adulthood- ness, 40, 165, 204, 215

age, 12, 14, 15, 20, 21, 27, 29, 32, 33, 40, 41, 51, 52, 53, 55, 56, 58, 62, 63, 64, 69, 70, 71, 72, 73, 74, 78, 79, 81, 89, 97, 98, 115, 116, 117, 122, 125, 127, 133, 150, 151, 152, 153, 154, 155, 156, 158, 160, 163, 170, 177, 178, 202, 203, 204, 206, 211, 212, 213, 221, 222, 223, 225, 227, 232, 234, 250, 251, 253, 255, 259, 260, 261, 262, 264, 266, 267, 271, 272, 273, 279, 280, 281, 282, 283, 284, 285, 287, 289, 290, 291, 292, 293, 294, 295

ageing, 14, 150, 151, 202, 221, 223, 228, 229, 231, 234, 237, 247, 249, 250, 255, 258

agency, 58, 88, 103, 122, 123

archive, 12, 13, 63, 163, 176, 177, 178, 179, 180, 183, 184, 249

audiovisual, 12, 175, 176, 177, 184, 194, 196

blog, 28, 64, 91, 105, 112, 114, 116, 140, 151, 167, 178, 193, 196, 305

boomer 21, 22, 23, 24, 29, 51, 60, 83, 85, 86, 88, 97, 106

broadband,
    society, 8, 14, 247, 255, 256, 308
    connection, 138, 210

children, 10, 11, 15, 19, 24, 26, 27, 29, 33, 48, 52, 59, 71, 121, 122, 123, 124, 125, 127, 129, 130, 134, 158, 163, 165, 169, 202, 204, 205, 206, 208, 209, 210, 211, 212, 213, 214, 215, 216, 217, 289, 290, 295, 297, 288, 299, 300, 302, 305

communication,
    devices, 234
    interpersonal, 30, 302
    technologies, 15, 221, 224, 256, 257, 295, 303

community, 13, 20, 44, 52, 92, 104, 113, 123, 125, 129, 141, 145, 154, 163, 187, 191, 192, 193, 194, 195, 196, 197, 198, 199, 223, 236, 283, 293, 294

connection,
    broadband, 134, 210

consumption, 10, 12, 22, 23, 24, 26, 30, 47, 51, 53, 55, 69, 98, 133, 134, 135, 136, 137, 142, 145, 158, 160, 161, 166, 190, 191, 295, 296, 302, 305 computer, 8, 15, 47, 107, 121, 124, 139, 151, 192, 193, 201, 209, 210, 211, 212, 213, 214, 215, 225, 226, 227, 251, 252, 253, 274, 277, 289, 290, 291, 293, 294, 295, 296, 297, 298, 299, 300, 301, 302, 303

content analysis, 113, 125, 159, 192, 208

countries,
    estonia, 11, 78, 121, 124, 129, 305
    finland, 70, 72, 73, 74, 78, 79, 84, 87, 88, 89, 305
    italy, 19, 21, 22, 29, 30, 33, 34, 72, 81, 87, 92, 115, 151, 154, 156, 157, 159, 170, 191, 207, 244, 249, 251, 277, 281, 285, 305, 311
    slovenia, 14, 221, 289, 290, 292, 293, 305
    turkey, 11, 131, 133, 134, 135, 139, 140, 141, 142, 143, 144, 305

culture,
    media, 11, 13, 300, 305
    peer, 7, 11, 121, 123, 124, 130, 132
    youth, 90, 98, 135, 142, 145, 194

digital,
    digitalization, 63, 174, 175, 180
    divide, 12, 14, 97, 122, 150, 165, 206, 272
    generation, 10, 52, 59, 121, 122, 130, 207, 209, 210, 212, 213, 214, 215, 216,
    immigrant, 14, 60, 206, 216, 261, 274
    native, 13, 60, 97, 150, 151, 165, 166, 169, 201, 203, 206, 207, 208, 209, 215, 216, 217, 259, 305

domestication, 10, 15, 216, 289, 290, 296, 297, 299

education(al), 14, 19, 24, 51, 52, 53, 60, 78, 130, 134, 144, 163, 195, 213, 214, 215, 216, 230, 251, 262, 263, 264, 265, 268, 269, 270, 272, 281, 292, 293, 294, 295, 297, 298, 302

ethnography, 113, 312, 311

eye-tracking, 14, 260, 261, 262, 263, 264, 269, 270, 271, 272

estonia, 11, 78, 121, 124, 129, 305

family, 20, 26, 53, 54, 70, 100, 101, 102, 127, 138, 139, 149, 167, 206, 215, 227, 229, 250, 254, 282, 289, 290, 292, 295, 296, 297, 298, 299, 300, 301, 302, 314

fan(dom), 8, 13, 64, 128, 141, 142, 189, 190, 191, 192, 193, 194, 195, 196, 197, 198, 201, 200, 305

finland, 70, 72, 73, 74, 78, 79, 84, 87, 88, 89, 305

focus group, 58, 99, 103, 137, 138, 234, 235, 236, 237, 239, 242, 249, 251

generation(al),
    broadband, 10
    definition of, 9, 62
    digital, 10, 52, 59, 121, 122, 130, 207, 209, 210, 212, 213, 214, 215, 216
    gap, 8, 12, 97, 159, 165, 170
    in-between, 165, 166
    internet, 10, 19, 28, 60, 64, 122
    marketing, 23, 24, 59, 62
    net, 23, 51, 60, 61, 121, 150, 210

    old, 60, 208
    problem of, 7, 10, 37, 39, 44, 69, 206
    young, 7, 9, 37, 38, 41, 46, 218, 296, 298
    x, 23, 114, 116
    y, 10, 99, 114, 116, 307
    web, 60, 64

gender, 14, 24, 31, 53, 84, 115, 116, 121, 125, 127, 128, 129, 137, 151, 161, 165, 175, 195, 206, 229, 266, 265, 266, 268, 276, 277, 288, 297, 300, 301, 302, 303, 316, 319

history,
    media, 7, 10, 39, 69, 89, 117, 311

household, 8, 15, 26, 30, 48, 85, 97, 133, 293, 294, 295, 296, 297, 299, 302

ICTs, 7, 8, 9, 10, 14, 51, 52, 53, 60, 61, 62, 63, 64, 65, 97, 98, 223, 222, 225, 226, 227, 228, 229, 230, 231, 249, 252, 253, 254, 258, 301, 303, 306, 311

innovation, 13, 21, 23, 24, 28, 54, 60, 62, 234, 235, 246, 313

internet,
    generation, 10, 19, 28, 60, 64, 122
    use(r), 14, 31, 112, 122, 140 143, 191, 218, 264, 265, 266, 268, 269, 271, 274, 275, 302, 303, 312

interview, 14, 15, 32, 99, 102, 113, 114, 127, 132, 137, 138, 140, 141, 142, 143, 144, 145 151, 162, 168, 192, 212, 216, 236, 245, 251, 252, 258, 285, 286, 287, 288, 291, 300, 301, 303, 305, 310

italy, 19, 21, 22, 29, 30, 33, 34, 72, 81, 87, 92, 115, 151, 154, 156, 157, 159, 170, 191, 207, 244, 249, 251, 277, 281, 285, 305, 311

marketing,
    generational, 23, 24, 59, 62

media,
    consumption, 10, 30, 51, 298, 301, 311,
    history, 7, 10, 39, 69, 89, 117, 310
    social, 11, 15, 97, 99, 100, 101, 103, 104, 105, 189, 191, 201, 281, 283, 286, 311, 316
    use(r), 53, 61, 93, 117, 217, 295, 299

memory, 8, 12, 13, 27, 28, 30, 58, 62, 63, 65, 175, 180, 181, 183, 205, 211, 227, 229, 263

methodology,
    content analysis, 113, 125, 159, 192, 208
    ethnography, 113, 312, 311
    eye-tracking, 14, 260, 261, 262, 263, 264, 269, 270, 271, 272
    focus group, 58, 99, 103, 137, 138, 234, 235, 236, 237, 239, 242, 249, 251
    interview, 14, 15, 32, 99, 102, 113, 114, 127, 132, 137, 138, 140, 141, 142, 143, 144, 145 151, 162, 168, 192, 212, 216, 236, 245, 251, 252, 258, 285, 286, 287, 288, 291, 300, 301, 303, 305, 310
    participant observation, 15, 99, 194, 283
    qualitative, 58, 112, 113, 115, 116, 178, 187, 194, 223, 243, 255, 301, 311,
    quali-quantitative , 249, 252, 253
    quantitative, 12, 116, 118, 137, 151, 156, 182, 183, 195, 154, 257, 312

millennials, 61, 150

mobile,
- phone, 13, 30, 64, 98, 99, 102, 104, 133, 136, 138, 140, 160, 162, 171, 205, 209, 211, 212, 213, 216, 217, 218, 247, 250, 251, 252, 254, 298, 299, 301, 312
- telecommunication, 102, 134

networks,
- social, 7, 8, 10, 12, 14, 15, 28, 30, 33, 64, 101, 102, 105, 121, 134, 138, 139, 140, 141, 149, 150, 151, 154, 155, 156, 157, 168, 178, 180, 214, 216, 217, 253, 281, 282, 283, 284, 285, 286, 290, 291, 312, 313

old,
- people, 225
- generation, 60, 208

parent(al), 11, 15, 19, 22, 24, 26, 30, 48, 59, 125, 126, 134, 136, 138, 162, 164, 166, 167, 168, 169, 206, 206, 207, 208, 213, 215, 216, 217, 218, 219, 295, 296, 298, 300, 301, 302, 303, 304, 306, 307

participant observation, 15, 99, 194, 283

people,
- old, 227
- young, 7, 10, 11, 12, 21, 22, 23, 30, 31, 33, 41, 72, 97, 98, 99, 106, 110, 121, 122, 124, 125, 129, 133, 134, 136, 137, 138, 139, 140, 141, 142, 143, 144, 145, 146, 148, 159, 160, 161, 162, 164, 165, 168, 169, 175, 185, 189, 192, 201, 206, 209, 210, 211, 213, 214, 216, 217, 218, 219, 220, 255, 286, 291, 296, 298, 299, 300, 301, 302, 306, 307, 309, 314

political
- action, 38, 61

privacy, 23, 138, 142, 230, 256, 295

private,
- sphere, 15, 138, 296, 298, 301, 302, 304, 308

public,
- sphere, 33, 64, 83, 87, 88, 139, 141, 144, 145, 314

semiotics, 175, 176, 178, 311

slovenia, 14, 221, 289, 290, 292, 293, 305

social,
- action, 104, 105, 201
- networks, 7, 8, 10, 12, 14, 15, 28, 30, 33, 64, 101, 102, 105, 121, 134, 138, 139, 140, 141, 149, 150, 151, 154, 155, 156, 157, 168, 178, 180, 214, 216, 217, 253, 281, 282, 283, 284, 285, 286, 290, 291, 312, 313
- practice, 11, 56, 100, 104, 105, 190
- relationship, 12, 141, 144, 171, 197, 215, 216, 218, 282, 302
- theory, 39

social networking site (SNS), 11, 12, 99, 110, 111, 122, 124, 125, 129, 138, 140, 141, 142, 150, 187, 189, 283, 284

society,
    broadband, 8, 249, 258, 312
sphere,
    private, 15, 138, 296, 298, 301, 302, 304, 308
    public, 33, 64, 83, 87, 88, 139, 141, 144, 145, 314
teenage(r), 33, 98, 122, 170, 193
telecommunication
    mobile, 102, 312
television, 13, 26, 27, 29, 32, 47, 70, 83, 84, 85, 87, 90, 91, 117, 162, 175, 176, 179, 189, 190, 191, 192, 193, 198, 201, 217, 227, 301, 302, 307, 308, 309, 313
theory,
    domestication, 10, 15, 220, 295, 296, 302, 303, 305
    semiotics, 175, 176, 178, 311
    social, 39
turkey, 11, 131, 133, 134, 135, 139, 140, 141, 142, 143, 144, 305
user generated content (UGC), 110, 112, 121, 177, 180, 286
young,
    adult, 9, 10, 13, 37, 152, 191, 193, 197, 230, 231, 254, 268, 288
    generation, 7, 9, 37, 38, 41, 46, 218, 296, 298
    people, 7, 10, 11, 12, 21, 22, 23, 30, 31, 33, 41, 72, 97, 98, 99, 106, 110, 121, 122, 124, 125, 129, 133, 134, 136, 137, 138, 139, 140, 141, 142, 143, 144, 145, 146, 148, 159, 160, 161, 162, 164, 165, 168, 169, 175, 185, 189, 192, 201, 206, 209, 210, 211, 213, 214, 216, 217, 218, 219, 220, 255, 286, 291, 296, 298, 299, 300, 301, 302, 306, 307, 309, 314
    youth, 28, 29, 32 33, 37, 40, 41, 51, 53, 59, 60, 70, 71, 72, 73, 76, 78, 80, 81, 82, 83, 85, 89, 90, 98, 109, 122, 126, 129, 131, 133, 134, 135, 136, 137, 138, 141, 142, 143, 144, 160, 165, 166, 191, 196, 251, 255, 298
web,
    2.0, 11, 33, 47, 100, 110, 111, 149, 179, 184, 189, 196, 197, 198
    generation, 60, 64
    site, 12, 14, 124, 125, 129, 140, 141, 142, 143, 144, 151, 177, 183, 215, 263, 264, 265, 267, 269, 273, 281, 283
we sense, 7, 11, 31, 32, 34, 40, 43, 48, 53, 55, 57, 61, 109, 110, 116, 117, 118

Participation in Broadband Society

Edited by Leopoldina Fortunati / Julian Gebhardt / Jane Vincent

This series publishes peer-reviewed monographs and edited volumes by internationally renowed scholars in the field of the 'social use of information and communication technologies (mass media included)', 'communication studies' and 'science and technology studies'. It provides an editorial space specifically dedicated to the collection of work that integrates new research regarding theoretical discourse, methodologies and studies from multiple disciplines such as sociology, anthropology, psychology, geography, linguistics, information science, engeneering and more.

The editors particularly welcome texts elaborating new theories, original methodological approaches and challenges to existing knowledge. Proposals aimed at scholars, professionals and operators working in the diverse field of participation in broadband society are invited from all disciplines.

Vol.	1	Leopoldina Fortunati / Jane Vincent / Julian Gebhardt / Andraz Petrovčič / Olga Vershinskaya (eds.): Interacting with Broadband Society. 2010.
Vol.	2	Julian Gebhardt / Hajo Greif / Lilia Raycheva / Claire Lobet-Maris / Amparo Lasen (eds.): Experiencing Broadband Society. 2010.
Vol.	3	Leslie Haddon (ed.): The Contemporary Internet. National and Cross-National Studies. 2010.
Vol.	4	Hajo Greif / Larissa Hjorth / Amparo Lasén / Claire Lobet-Maris (eds.): Cultures of Participation. Media Practices, Politics and Literacy. 2011.
Vol.	5	Fausto Colombo / Leopoldina Fortunati (eds.): Broadband Society and Generational Changes. 2011.

www.peterlang.de

www.ingramcontent.com/pod-product-compliance
Ingram Content Group UK Ltd.
Pitfield, Milton Keynes, MK11 3LW, UK
UKHW021822140426
5217IPUK00004B/49